International Workshop on Fission Product Yields (FPY 2019)

EPJ Web of Conferences Volume 242 (2020)

Santa Fe, New Mexico, USA
30 September - 4 October 2019

Editor:

Toshihiko Kawano

ISBN: 978-1-7138-1920-2

Printed from e-media with permission by:

Curran Associates, Inc.
57 Morehouse Lane
Red Hook, NY 12571

Some format issues inherent in the e-media version may also appear in this print version.

For additional information, please contact EDP Sciences – Web of Conferences at the address below.

EDP Sciences – Web of Conferences
17, Avenue du Hoggar
Parc d'Activité de Courtabœuf
BP 112
F-91944 Les Ulis Cedex A
France

Phone: +33 (0) 1 69 18 75 75
Fax: +33 (0) 1 69 28 84 91

contact-edps@webofconferences.org

Additional copies of this publication are available from:

Curran Associates, Inc.
57 Morehouse Lane
Red Hook, NY 12571 USA
Phone: 845-758-0400
Fax: 845-758-2633
Email: curran@proceedings.com
Web: www.proceedings.com

TABLE OF CONTENTS

edp sciences

EPJ Web of Conferences

Statement of Peer review

In submitting conference proceedings to EPJ Web of Conferences, I certify to the Publisher that I adhere to the **Policy on Publishing Integrity** of the journal in order to safeguard good scientific practice in publishing.

1. All articles have been subjected to peer review administered by the proceedings editors.
2. Reviews have been conducted by expert referees, who have been requested to provide unbiased and constructive comments aimed, whenever possible, at improving the work.
3. Proceedings editors have taken all reasonable steps to ensure the quality of the materials they publish and their decision to accept or reject a paper for publication has been based only on the merits of the work and the relevance to the journal.

Title, date and place of the conference

International Workshop on Fission Product Yields
Inn and Spa at Loretto, Santa Fe, New Mexico, USA
Sep.30 - Oct.4, 2019

Proceedings editor(s): Toshihiko Kawano

Sep. 15, 2020

17, avenue du Hoggar - PA de Courtabœuf – BP 112 - 91944 Les Ulis Cedex A (France)
Tél. : 33 (0)1 69 18 75 75 - Fax : 33(0)1 69 07 45 17 – www.edpsciences.org

EPJ Web of Conferences **242**, 01001 (2020) https://doi.org/10.1051/epjconf/202024201001
FPY 2019

Measurements of Fission Products Yields with the LOHENGRIN mass spectrometer at ILL

A. Chebboubi[1,a], S. Julien-Laferrière[1], J. Nicholson[1], G. Kessedjian[2], O. Serot[1], A. Blanc[3], D. Bernard[1], H. Faust[3], Y.H. Kim[3], U. Köster[3], A. Letourneau[4], O. Litaize[1], O. Méplan[2], P. Mutti[3], M. Rapala[4], M. Ramdhane[2], and C. Sage[2]

[1]*CEA, DEN, DER, SPRC, Cadarache, Physics Studies Laboratory, F-13108 Saint-Paul-lès-Durance, France*
[2]*LPSC, Université Grenoble-Alpes, CNRS/IN2P3, F-38026 Grenoble Cedex, France*
[3]*Institut Laue-Langevin, F-38042 Grenoble Cedex 9, France*
[4]*IRFU, CEA,Université Paris-Saclay, 91191 Gif-sur-Yvette, France*

Abstract. The CEA in collaboration with ILL and LPSC has developed a measurement program on symmetric and heavy mass fission product distributions. The combination of measurements with ionisation chamber and Ge detectors is necessary to describe precisely the heavy fission product region in mass and charge. Recently, new measurements of fission yields and kinetic energy distributions, for different fissioning systems $\left({}^{233,235}\mathrm{U}(n_{th}, f), {}^{241}\mathrm{Am}(2n_{th}, f) \text{ and } {}^{239,241}\mathrm{Pu}(n_{th}, f)\right)$, were performed with recoil spectrometer LOHENGRIN. The focus has been done on the self-normalization of the data to provide new absolute measurements, independently of any libraries along with the experimental covariance matrix. To reach precise measurements, a new experimental procedure was developed along with a new analysis method.

1 Introduction

The mass and isotopic yields of the fission products have a direct influence on the amount of neutron poisons that limit the fuel burnup and on the evaluation of the residual power of the reactor after shutdown [1, 2]. The fission yields of the plutonium nuclides are also mandatory for the studies on the fuel multi-recycling [3]. In order to significantly improve the precision of nuclear data, more and more fundamental fission models are used in the evaluation processing. Therefore, tests of fission models become a central issue to achieve coherent libraries of nuclear data. In this framework, an important investigation in the experimental limits of facilities is required to provide complete sets of data with their coherent variance-covariance matrices. In the past with the LOHENGRIN spectrometer [4] of the Institut Laue-Langevin (ILL), priority has been given for the studies in the light fission fragment mass range. This approach was due to the opportunity to extract the mass and charge distribution using a double Frisch grid ionization chamber (IC). Since 2010, the CEA in collaboration with ILL and the LPSC study the fission products properties by combining IC measurements with γ spectroscopy using High Purity Germanium (HPGe) detectors. These studies are focused on the heavy mass region. Along he years, several fissioning systems such as ${}^{239}\mathrm{Pu}(n_{th}, f)$ [5, 6], ${}^{241}\mathrm{Am}(2n_{th}, f)$ [7],

[a]e-mail: abdelhazize.chebboubi@cea.fr

^{235}U(n_{th}, f) [8], ^{233}U(n_{th}, f) [9] and ^{241}Pu(n_{th}, f) [10] were measured. In this article, a focus on the mass yield measurements is made. Then some results on isotopic yields and complementary observables are shown.

2 Measurements of fission product mass yields $Y(A)$

The measurement of $Y(A)$ were done at the LOHENGRIN recoil separator of the ILL. The different parts of the LOHENGRIN spectrometer are shown on Fig. 1 (left). The fissile isotope target is placed near the core of the reactor under a thermal neutron flux about 5×10^{14} n.cm^{-2}.s^{-1}. The combination of a magnetic and electric fields allows a separation of the produced fission products according to their ratios mass A over ionic charge q and kinetic energy E_k over ionic charge. Finally a refocusing magnet [11] permits to increase the particle density at the focal position 2. To remove the mass degeneracy an ionization chamber (at the focal position 1) is used to measure E_k. Figure 1 (right) shows the energy loss inside the first part of the IC as function of the total energy loss inside the IC. Each spot is then associated to a mass number.

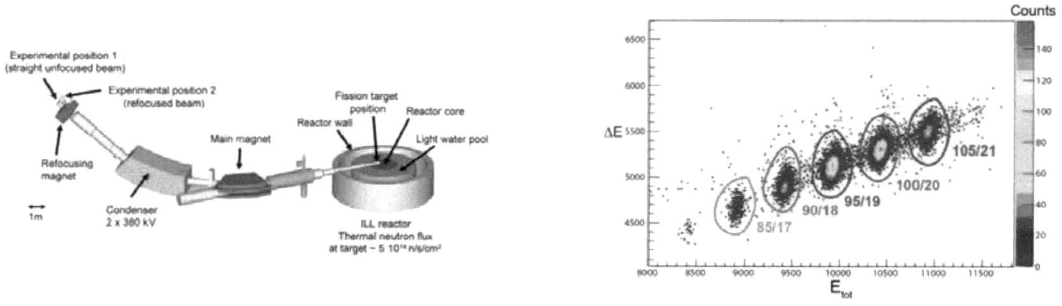

Figure 1. Scheme of the LOHENGRIN spectrometer (left). Plot of the energy loss ΔE in the first part of the IC and the total energy loss E_{tot} inside the IC. Each spot corresponds to a kinetic energy, which refers to the fission product mass (right). The mentionned numbers correspond to the ratio A/q.

By definition, the relative mass yield $\mathcal{N}(A)$ is written :

$$\mathcal{N}(A) = \sum_q \int_{E_k} \mathcal{N}(A, q, E_k) dE_k \tag{1}$$

with $\mathcal{N}(A, q, E_k)$ the count rate for a given ionic charge and kinetic energy selected by the LOHENGRIN spectrometer. However because of the finite energy resolution of the spectrometer, measurement by energy bins must be done. Also because of the high flux, the fissile matter evolves quite rapidly with time, typically 20 days of irradiation for a ^{241}Pu sample of 100 μg.cm^{-2}. In order to measure the maximum number of masses (during a single experimental campaign), some approximations are then made:

$$\mathcal{N}(A, q, E_k) = \frac{N_{ion}(A, q, E_k \pm \Delta E_k, t)}{BU(t) \times \Delta E_k} \tag{2}$$

with $N_{ion}(A, q, E_k \pm \Delta E_k, t)$ the count rate extracted from the IC at a given kinetic energy, ionic charge and time. The kinetic energy resolution depends on the aperture at the exit of the condenser. Here the kinetic energy resolution is: $\Delta E_k = 0.005 \times E_k$, which corresponds to an aperture of 3.24 cm[1]. If

[1] In the following, E_k refers to $E_k \pm \Delta E_k$

we suppose that the ionic charge and the kinetic energy are two independent parameters, then eq. (1) becomes:

$$\mathcal{N}(A) = \frac{\sum_{E_k} \mathcal{N}(A, q_m, E_k)}{P(q_m, E_k^m)}$$
$$P(q_m, E_k^m) = \frac{\mathcal{N}(A, q_m, E_k^m)}{\sum_q \mathcal{N}(A, q, E_k^m)} \tag{3}$$

with $P(q_m, E_k^m)$ the probability to measure the ionic charge q_m at a given kinetic energy E_k^m. In other words, a kinetic energy distribution measurement is made at the matching ionic charge $q = q_m$ and a ionic charge distribution measurement is made at the matching kinetic energy $E_k = E_k^m$. The underlying hypothesis is that the kinetic energy distribution does not depend on the ionic charge (and respectively) which is not true as demonstrated in Fig. 2. Therefore a correction of the eq. (3) must be done in order to correctly estimate the relative mass yield $\mathcal{N}(A)$. Concretely, average and width of the kinetic energy distributions are extracted as function of the ionic charge. A linear fit is then performed:

$$\overline{E_k}(q) = aq + b$$
$$\sigma_{E_k}(q) = cq + d \tag{4}$$

Then the correction due to the (E_k, q) correlation is performed through the ionic charge distribution. This correction is made in order to answer to the following question: what would be the weight of each $P(q, E_k^m)$ if the measurement was made at a different kinetic energy? $P(q_m, E_k^m)$ is corrected through:

$$P^{corr}(q_m, E_k^m) = \frac{P(q_m, E_k^m) \exp\left(-\frac{\left(\overline{E_k}(q_m) - E_k^m\right)^2}{\sigma_{E_k}(q_m)}\right)}{\sum_q P(q, E_k^m) \exp\left(-\frac{\left(\overline{E_k}(q) - E_k^m\right)^2}{\sigma_{E_k}(q)}\right)} \tag{5}$$

In this expression, the kinetic energy distribution is assumed Gaussian, which is a good approximation when using a thin target (typically under 100 μg.cm^{-2}). With this new procedure, a relative mass yield can be estimated for different ionic charges q:

$$\mathcal{N}(A|q) = \frac{\sum_{E_k} \mathcal{N}(A, q, E_k)}{P^{corr}(q, E_k^m)} \tag{6}$$

The mean value is derived using the following formula:

$$\overline{\mathcal{N}}(A) = \left(\sum_{i,j}^{n,n} \left(C^{-1}\right)_{i,j}\right)^{-1} \left(\sum_{i,j}^{n,n} \left(C^{-1}\right)_{i,j} \mathcal{N}(A|q_i)\right) \tag{7}$$

with C the covariance matrix between each $\mathcal{N}(A, q_i)$ and n the number of kinetic energy distributions. The compatibility between each $\mathcal{N}(A, q_i)$ is achieved through a generalized χ^2 test with a confidence level of 90%. If the test is not passed, an uncertainty is incrementally added on the diagonal value of C [12]. This additional uncertainty can be interpreted as the limits of the approximations made (linear fit used in eq. (4)) and as the consequence of the LOHENGRIN stability. Figure 3 shows the impact of the correlation (E_k, q). Both the uncertainty and the bias are reduced. In order to produce independent

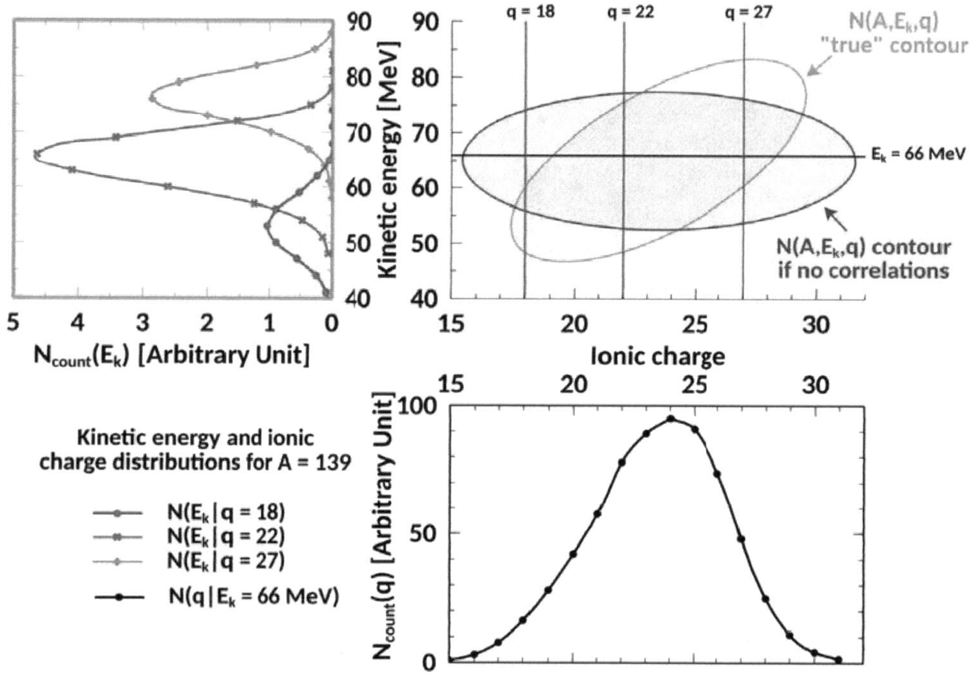

Figure 2. Dependence of the kinetic energy distribution as function of the ionic charge. The $\mathcal{N}(A, q, E_k)$ contour is then rotated thanks to the correlation between (E_k, q). Data are coming from the study on the $^{241}\text{Pu}(n_{th}, f)$ reaction.

data, the whole heavy (or light peak) must be measured (at least 99.5 %). Then the absolute mass yields $Y(A)$ are written:

$$Y(A) = \frac{\mathcal{N}(A)}{\sum_{A \in \text{Heavy}} \mathcal{N}(A)} \tag{8}$$

With this procedure the final uncertainty is of the order of 3% for the highly produced masses ($Y(A) >$ 1%). The covariance matrix can also be derived reflecting the raw-data analysis process.

3 Measurements of fission product isotopic yields $Y(A, Z)$

To assess the nuclear charge of the fission products (in the heavy mass region), γ spectroscopy is used on the LOHENGRIN recoil spectrometer. A movable tape is placed inside a vacuum chamber surrounded by two High Purity Germanium (HPGe) clovers (with 4 Ge crystals each). The setup is positioned at the focal position 2. The procedure for measuring isotopic yields is described hereafter. First kinetic energy distributions and ionic charge distribution are measured with the IC (placed at the focal position 1). Then the ionic charge distribution is also measured with the HPGe and allows to determine $N(Z|A)$ which corresponds to the relative isotopic yields. Finally the isotopic yields is

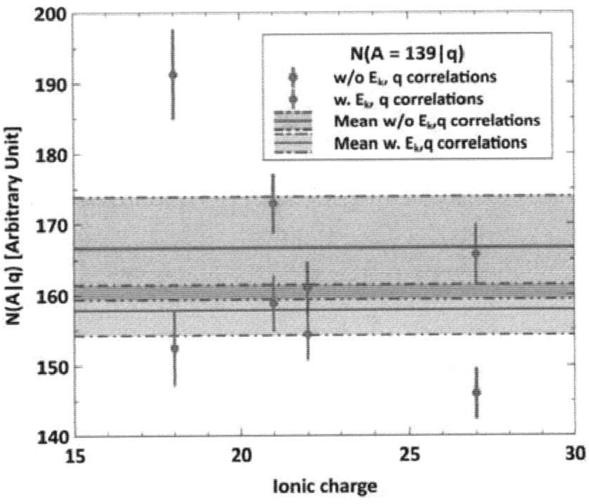

Figure 3. Estimation of the relative mass yield for $A = 139$ from the ^{241}Pu(n_{th}, f) reaction. Taking into account the correlation (E_k, q) permits to reduce the uncertainty and the bias introduced by the classical procedure to estimate the fission mass yields on LOHENGRIN.

given by:

$$Y(A, Z) = N(Z|A) \times k_{139}$$
$$k_{139} = Y(A = 139) / \sum_Z N(Z|A = 139) \quad (9)$$

In this method, the mass $A = 139$ is supposed to be completely measured through γ spectroscopy. Moreover, some corrections [5, 13] are needeed to transform the γ count rate $N_\gamma(Z, q, E_k^m)$ to the relative isotopic yield N(Z|A).

Here a focus on the impact of γ intensity I_γ is made. Indeed in the case of the ^{136}Te, two different sets of γ rays are in agreement. As shown on Fig. 4 (left), to assess the production rate of ^{136}I, coming from fission, the β^- decay of ^{136}Te to ^{136}I must be corrected. A negative solution is found for the solution 2 as shown on Fig. 4 (right). In other words, the relative intensity between those γ are not consistent. This example shows how the isotopic yields are sensitive to the nuclear structure parameters.

4 Other observables of interest to study the fission process

Thanks to the versatility of the LOHENGRIN recoil spectrometer, indirect fission observables can be studied. For instance using the mass yield measurements, the mean kinetic energy as function of mass can be derived [14]. This quantity can be compared with other measurements and can be used by different models such as FIFRELIN [15, 16] to validate the input data and the underlying models. The study of the symmetry region can give insight on the Brosa fission modes [17] by looking at the kinetic energy distributions [18]. The local odd-even effect [13] give some information on the pairs breaking but also on the neutron emission process. An experimental program on isomeric ratios (IRs) measurements was developed since 2012 [19]. Especially a focus on the evolution of IR as function

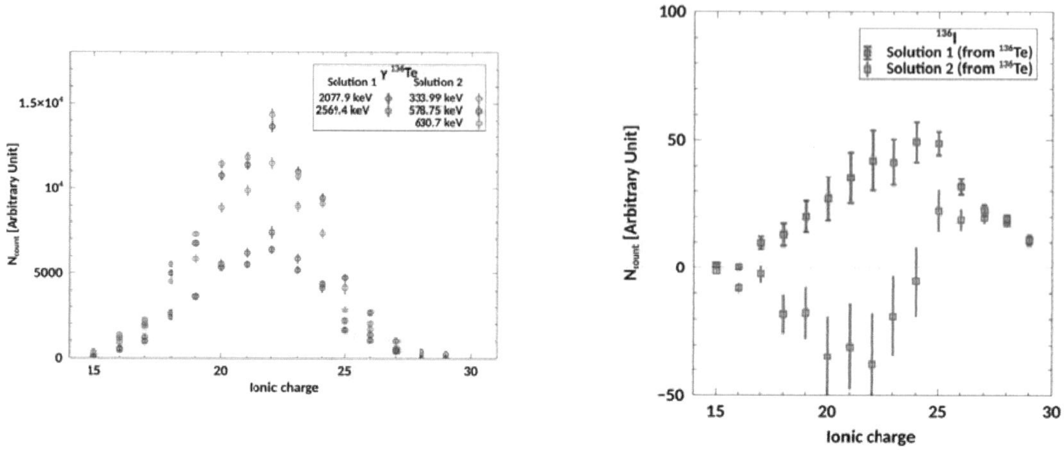

Figure 4. Count rate extracted from γ spectrum (left). Tow different sets of γ-rays (solution 1 and solution 2) are in agreement within each other. Correction from β^- decay of ^{136}Te to ^{136}I give rise to two contradictory solutions (right). Data are coming from the studies of the ^{241}Pu(n_{th}, f) reaction.

of fission product kinetic energy was made [20, 21]. Using de-excitation code such as FIFRELIN, an angular momentum distribution can be derived. These data can also be used to test the level density or γ-strength function models as well as the coherency of the nuclear level schemes. Finally looking at the structures in the kinetic energy distributions coming from a high quality thin target allows to make an indirect measurement of the neutron emission process for a given mass.

5 Conclusion

This work presents the new procedure developed on the LOHENGRIN recoil spectrometer in order to measure mass and isotopic yields. Taking into account the correlation (E_k, q) permits to improve the accuracy and uncertainties of the measurements. The strong sensibility of isotopic yield on nuclear structure parameters was highlighted. Thanks to the performance of the spectrometer, indirect measurement of observable of interest are currently studied in order to test fission models or nuclear structure models.

6 Acknowledgment

This work was supported by IN2P3, by the University of Grenoble Alpes and by "le défi NEEDS". The authors are grateful for the support of the ILL and all the staff involved from CEA-Cadarache and LPSC.

References

[1] V. Vallet *et al.*, *Proc. Physor 2014*, 2014-003 (2015)
[2] A. Tsilanizara *et al.*, *Ann. Nucl. Energy* **90**, 62-70 (2016)
[3] M. Tiphine *et al.*, *Proc. Global 2015*, 5327 (2015)
[4] P. Armbruster *et al.*, *Nucl. Instrum. Methods* **139**, 213 (1976)

[5] A. Bail *et al.*, *Phys. Rev. C*, **84**, 034605 (2011)
[6] Y.K. Gupta *et al.*, *Phys. Rev. C*, **96**, 014608 (2017)
[7] C. Amouroux *et al.*, *EPJ Web Conf.* **62**, 06002 (2013)
[8] O. Serot *et al.*, *Nucl. Data Sheets*, **119**, 320-323 (2014)
[9] A. Chebboubi *et al.*, *EPJ Web Conf.* **146**, 04021 (2017)
[10] S. Julien-Laferière *et al.*, *ND2019 Conference*, to be published
[11] G. Fioni *et al.*, *Nucl. Instrum. Meth. A* **332**, 175 (1993)
[12] S. Julien-Laferière *et al.*, *EPJ Nucl. Sci. Technol.* **4**, 25 (2018)
[13] S. Julien-Laferière *et al.*, to be submitted
[14] S. Julien-Laferière *et al.*, *EPJ Web Conf.* **169**, 00008 (2018)
[15] O. Litaize and O. Serot, *Phys. Rev. C* **82**, 054616 (2010)
[16] O. Litaize *et al.*, *EPJ A* **51**, 1 (2015)
[17] U. Brosa *et al.*, *Phys. Rep.* **197**, (4) 167-262 (1990)
[18] A. Chebboubi *et al.*, *EPJ Web Conf.* **111**, 08003 (2016)
[19] A. Chebboubi *et al.*, *EPJ Web Conf.* **111**, 08002 (2016)
[20] A. Chebboubi *et al.*, *Phys. Lett. B* **775**, 190-195 (2017)
[21] J. Nicholson *et al.*, *Theory-5 Conference*, to be published

Developments in New Measurements of Fission Cross-Sections, Fragment Yields, and Prompt and Quasi-Prompt Gammas for Nuclear Data Needs

Adam Hecht*, *Phoenix* Baldez, and *Mark* Wetzel
Department of Nuclear Engineering, University of New Mexico, Albuquerque, NM 87131, USA

Abstract. The University of New Mexico Fission Spectrometer was developed to measure fission product yield, as part of the LANL SPIDER collaboration. The spectrometer operates as an E-v detector to extract product mass event-by-event, with a time of flight region followed by an ionization chamber for kinetic energy measurements. By using the ionization chamber as a single-cathode/single-anode time projection chamber, stopping power and thus Z information is extracted, for coupled A and Z measurements. New work is being performed to add gamma ray detectors in the data stream, placed near the target region for prompt gammas and near the ionization chamber for quasiprompt (>50 ns) and later gammas, correlated with individual fission products. A stand-alone parallel plate ionization chamber (PPIC) is also being developed for fission tagging gamma ray data. The PPIC will also allow discrimination between charged particle out events and (n,n'γ), and discriminate between alpha emission and fission. Using layers in the PPIC, other targets can be measured simultaneously with a calibration target, giving relative fission cross sections. Past measurements with the spectrometer were performed at LANSCE and we plan to continue measurements there. The current work is supported by the NNSA Stewardship Science Academic Alliance.

1 Introduction

Nuclear data needs call for measurement of high-resolution fragment yields and energies for ^{239}Pu and ^{235}U fission. The Nuclear Energy Agency has begun to keep a list of high priority requests for data relating to nuclear energy applications [1]. This includes prompt gammas from ^{239}Pu and ^{235}U fission. Los Alamos National Laboratory (LANL) partners have also expressed the need for quasi-prompt gammas for studying fission daughter

* Corresponding author: hecht@unm.edu

isomers, while quasi-prompt gammas from fission parent isomers are important to understanding pre-fission nuclear shapes for theory.

As part of the SPectrometer for Ionization DEtermination in fission Research (SPIDER) effort, we have built an E-v fission product spectrometer at the University of New Mexico (UNM) for particle-by-particle fission mass yield data. With the data from within approximately 50 ns of fission we are acquiring quasi-prompt (>50ns) independent mass yields. With the energy detector, an ionization chamber (IC), which is wired as a time projection chamber, range information and thus Z information can also be extracted. With both A and Z we have N information. Thus we have coincident E, v, A, Z, and N information for each particle. Under a new Stewardship Science Academic Alliance grant, gamma detectors, including HPGe, will be placed near the target region for prompt coincidence gamma data and near the ionization chamber for quasi-prompt to delayed gammas. To increase fission tagging geometric efficiency, a stand alone ionization chamber with the fission target within the gas will be used with the gamma detectors. When run at the LANSCE neutron facility at LANL, the neutron TOF can be used in coincidence with the fission to develop the data sets as a function of neutron energy. The full spectrometer and the stand alone ionization chamber can be run in tandem in the same neutron beam in optimize beamtime. The full spectrometer has been developed and is the focus of this proceeding. Future work will include the standalone ionization chamber and gamma measurements.

2 Experiment and simulation

The UNM Fission Spectrometer is an E-v spectrometer which measures the kinetic energy and velocity of particles which, following $E=\frac{1}{2}mv^2$, can be combined to extract mass. A schematic diagram and a cut through view of the spectrometer is shown in Fig. 1. Some fission fragments pass through the time of flight (TOF) region and into the energy detector. The TOF region consists of a start and a stop time detector. Each timing detector each has a thin carbon foil (~50 μg/cm^2 for the discussed experiments) the fragment passes through and ejects electrons. The electrons are reflected by an electrostatic mirror to a microchannel plate detector placed to the side to not interrupt the travel of the fission fragments. The TOF path length is 50 cm and is operated in the range of 10^{-7} to 10^{-8} torr.

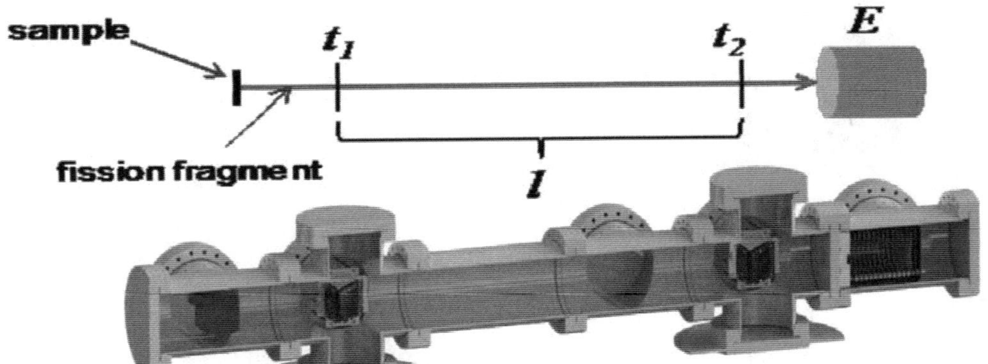

Fig. 1. (top) Schematic of time-of-flight region with target/sample region, t1 and t2 detectors and energy detection region. (bottom) Cut-through view of UNM Fission Spectrometer with target/sample region, t1, t2, and E detectors lined up with schematic.

Further along the path is the energy detector, a gas filled ionization chamber (IC) separated from the TOF region by a 200 nm (58 μg/cm^2) SiN window, Fig. 2. For the data discussed, the IC was operated with 70 torr of isobutane. IC detectors have better energy resolution for fission fragments than solid state detectors and suffer less cumulative damage. Gas cracking may occur with continued ionization and so new gas is cycled through the IC, with a refresh rate of 1 hour. The ionization chamber consists of an active region of about 8 cm between the cathode and Frisch grid, followed by an anode from which the energy pulse is read out. The entrance window consists of several 1 cm^2 SiN windows mounted in a stainless steel frame. This is mounted flush with the cathode annulus and is electrically connected with the cathode annulus to complete the cathode electrode. The cathode electrode is mounted on Teflon rather than directly to the vacuum flange to allow signal readout for range determination to extract Z. There is a series of guard rings between the cathode and the grounded Frisch grid to ramp down the voltage to produce a uniform electric field in the active region, which makes the energy pulse more uniform and allows the IC to be used as a time projection chamber for range determination.

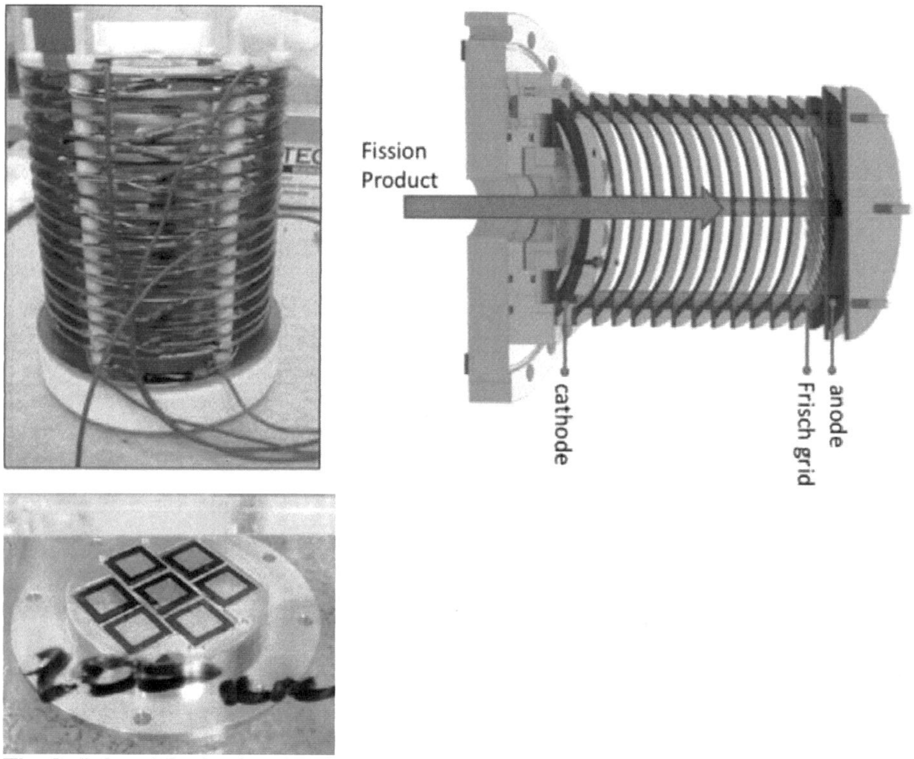

Fig. 2. (left top) Ionization chamber electrodes with cathode on top, the Teflon on the bottom is just structural, (right top) cut through of IC in use with fragment coming from the left, and (bottom) window frame with 7 SiN windows.

The energy of each particle in the IC is slightly lower than the energy of that particle in the TOF region due to the second TOF foil and the SiN window between the regions. The energy loss must be added back to correlate these two regions. We are pursuing beam calibrations, but for now this addback is performed using simulations. Comparison of SRIM-2013 [2,3] and MCNP6.2 [4] and our ^{252}Cf source measurements, Fig. 3, shows that MCNP overstates energy loss while SRIM understates the loss, making accurate addback using simulations difficult.

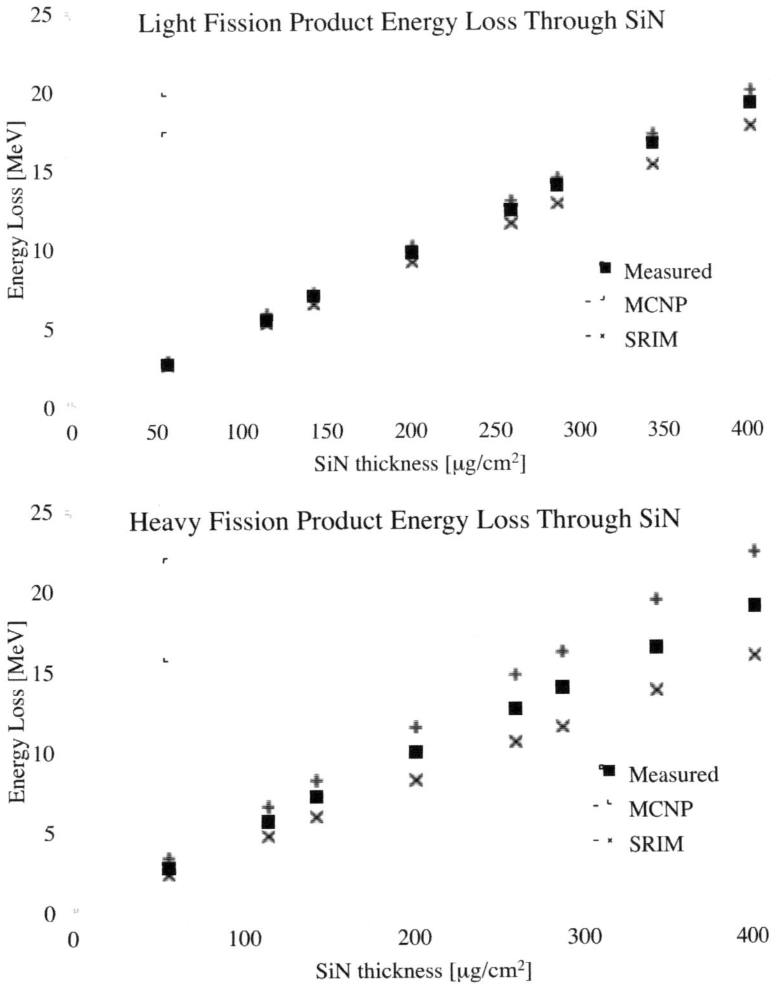

Fig. 3. Measured energy loss for ^{252}Cf light and heavy fission products through various thicknesses of SiN windows, measured with a Si surface barrier detector, compared with MCNP and SRIM predicted values. More details of this part of the work can be found in Baldez *et al.* [5].

For range determination, we follow the active cathode method of Sanami *et al.* [6] The fragment enters the IC and ionizes the isobutane gas. The electron cloud created begins drifting towards the Frisch grid, inducing a pulse on the cathode, and when it finally passes the Frisch grid it induces a pulse on the anode. The time difference between the beginning of the cathode and anode signals then relates to the range of the fragment in the IC gas. The Bethe formula relates stopping power to particle Z, but in this case the particle is not fully ionized. There is at least some proportionality expected and light and heavy fragments do show different ranges. To understand the relation between A, E, Z and range, SRIM simulations were performed and used to extract Z from the measured and extracted E, A, and range of each particle. SRIM predictions of range were close to the measured values for light fragments, about 1% high, though predictions were about 5% low for the heavy fragments.

3 Results

A plot of raw correlated TOF-IC values is presented in Fig. 4 (top). The heavy and light fragment concentrations are discernible at the lower and higher IC channels, respectively. The high amount of scatter may be due to the support wires for the carbon foils, the electrostatic mirror wires, and the many window frame edges. Using MCNP to determine the energy corrections, the masses were extracted, and are presented in Fig. 4 (bottom). Corrections with SRIM gave lower masses for the peaks. It is not clear why MCNP energy addbacks should work so well, since MCNP overestimates energy losses. Beam calibrations of energy loss are needed.

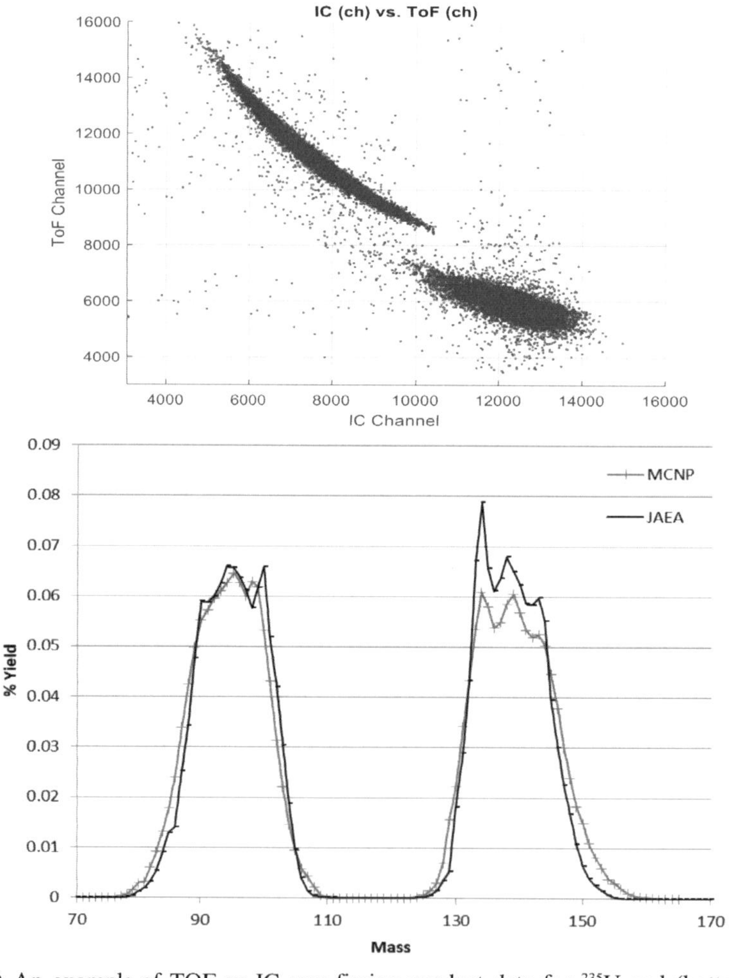

Fig. 4. (top) An example of TOF vs IC raw fission product data for ^{235}U and (bottom) extracted masses from this data using MCNP simulation based energy corrections compared with table values from JAEA.

Z information, extracted from the simulated range vs. A, Z, and E relations, was combined with A information to extract N for each particle, which is presented as Z vs. N and overlaid on stable nuclide positions in Fig. 5. The light fragment Z values are reasonable, though the heavy fragment Z values are too broadly spread, perhaps due to heavy fragment range simulations being too far from measurements. Calibration of range response and reconstruction of Z will be performed using x-rays from the ^{252}Cf source for identification.

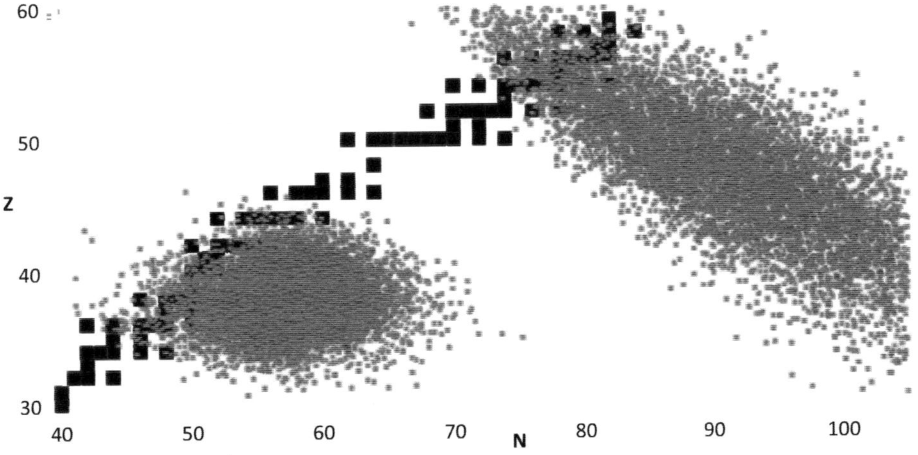

Fig. 5. Presentation of (N,Z) values extracted for each stable nuclides in black.

4 Further work

Further work is required. Beam calibrations were mentioned, to understand the pulse height response of the ionization chamber to different species at different energies, and to understand energy losses through the system. The range of particles in the IC must be better understood, which can also be studied with beam calibration. In addition, we will perform measurements of gamma rays in coincidence with the measured fission fragments, with both prompt gammas near the fission source/target and delayed gammas near the ionization chamber, with delayed being from 100 ns and on. To increase measurement efficiency, a stand alone ionization chamber, following Tovesson *et al.* [7], will be used for fission tagging gamma coincidences.

This work is partially supported by the NNSA Stewardship Science Academic Alliance, grant number DE-NA0003901. This leverages prior development work that was partially supported by DOE-NEUP, project number 13-4815.

References

[1] Nuclear Energy Agency, Nuclear Data High Priority Request List, available at <https://www.oecd-nea.org/dbdata/hprl>
[2] James F. Ziegler, M.D. Ziegler, J.P. Biersack, Nucl. Instr. Meth. **B 268**, 1818-1823 (2010).
[3] SRIM-2013 manual and code available from: <http://www.SRIM.org/>.
[4] J. Werner (Ed.), MCNP User's Manual – Code Version 6.2, LANL, Los Alamos, NM, USA, Rep. LA-UR-17-29981, 2017.
[5] P. Baldez, S. Fellows, R.E. Blakeley, M. Tanguay, M.L. Wetzel, A.A. Hecht, D. Mayorov, F. Tovesson, J. Winkebauer, Nucl. Instr. Meth. **B 456**, 142-147 (2019).
[6] T. Sanami, M. Hagiwara, T. Oishi, M. Baba, M. Takada, Nucl. Instr. Meth. **A 589**, 193-201 (2008).
[7] F. Tovesson, T.S. Hill, Phys. Rev. **C 75**, 034610 (2007).

Production yields of micro-second isomers populated in ^{252}Cf(SF)

Laurent Gaudefroy[1,*] and *Adeline* Ebran[1,**]

[1]CEA, DAM, DIF, F-91297 Arpajon, France

Abstract. A twin Frisch grid ionization chamber loaded with an ultra-thin ^{252}Cf sample is used in order to detect spontaneous fission events. The chamber is surrounded by 8 HPGe detectors allowing to perform γ-spectroscopic measurements. We are interested in the decay of isomeric states populated in fission fragments. In this contribution we report on preliminary results on production yields of isomeric states with half-lives ranging from few nano- up to few micro-seconds.

1 Introduction

Spontaneous fission sources offer an interesting possibility in order to study a large number of radioactive species that would otherwise require a particle or ion accelerator to be produced and excited. The large amount of unstable nuclei produced in the spontaneous fission process in general, such as that of ^{252}Cf in particular, however makes it difficult to disentangle the measured radiations. Selective experimental setups should therefore be developed in order to perform unambiguous association between radiations of interest and identified fission fragments.
In the present contribution we are interested in the delayed gamma-ray spectroscopy of fission fragments. Micro-second isomers might reflect profound structural changes in the wave functions of low-lying nuclear states therefore inhibiting transition rates between them. Studying the decay of these isomers can allow to locate where these structural changes occur in the Segré chart. With the help of nuclear models one can hope to pinpoint the origin of these changes in a particular region of the nuclide chart [1].

Short-lived nuclear isomers populated during the fission process can also provide interesting information about the fission process itself. Indeed the prompt particle, neutrons and gamma-rays, emission carry some information on the fragment shape configuration, excitation energy and angular momentum at scission. Naively, one expects a clear link between the spin-distribution in the primary fragments and the population of high-spin discrete states in the fragments produced after neutron-emission. In this works we therefore propose to measure the production yield of micro-second isomers in fission fragments. Comparison of these results with theoretical predictions should allow to constrain model parameters on spin-distribution in primary fission fragments.

*e-mail: laurent.gaudefroy@cea.fr
**e-mail: adeline.ebran@cea.fr

In the first part of this contribution, we present the experimental setup and the achieved selectivity allowing to perform gamma-spectroscopy of identified fission fragments. Then we present the procedure used in order to extract production yields of the populated short-lived isomeric states. Preliminary results are reported for ^{133}Te taken as an example. Perspectives are finally briefly depicted.

2 Experimental setup and selectivity

The experimental setup used in the present work consists of a twin Frisch grid ionization chamber loaded with an ultra-thin ^{252}Cf sample surrounded by 8 HPGe detectors. The principle of the measurement is as follows. Both fragments emerging from a spontaneous fission event are stopped in the chamber and their kinetic energy is measured. For cases where at least one fragment is produced in an excited long-lived (from few ns up to few μs) state, its gamma decay is measured using the HPGe detectors surrounding the ionization chamber. In this section more details on the just sketched experimental setup are reported as well as examples of the achieved selectivity.

Figure 1 shows the twin Frisch grid ionization chamber used in this work. It consists of two sets of axial chambers sharing a common cathode. All electrodes present an active diameter of 165 mm. The cathode-anode distance is 93 mm. Each anode is separated from the cathode by stainless-steel mesh serving as Frisch grids. The anode-grid distance is 13 mm. The chamber is filled with P-10 gas [Ar (90%) + CH_4 (10%)] at 900 mb. At this pressure the typical range for fission fragment amounts to 2 to 3 cm and the fragments are stopped within about 3 ns. The cathode voltage is set to -2100 V; the Frisch grid is grounded and the anode voltage is +1700 V.
The ionization chamber is loaded with a 15 kBq ultra-thin ^{252}Cf sample. The sample is supported by a 10 μg/cm^2 carbon backing. This ultra-thin sample has been prepared using the self-transfer technique [2, 3].
The ionization chamber is surrounded by 8 coaxial HPGe detectors the relative efficiency of which varies from 40% up to 80%. The HPGe detectors are vertically centered on the cathode (i.e. on the ^{252}Cf sample) and oriented in the direction normal to the symmetry

Figure 1. Picture of the twin Frisch grid ionization chamber used in this work. The central cathode support the ultra-thin ^{252}Cf sample. Distances between the electrodes are indicated.

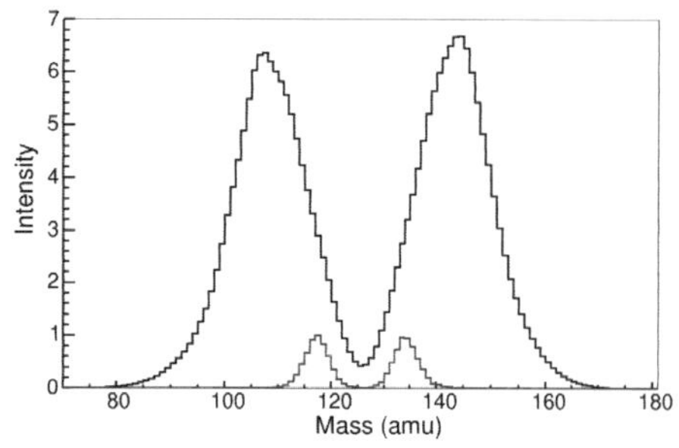

Figure 2. In black, preliminary post-neutron mass-distribution deduced from the measured kinetic energies in the ionization chamber. The blue curve shows the mass distribution gated by one of the gamma rays emitted in the decay of the 6^+ isomer at 1691 keV in ^{134}Te. Both curves are arbitrarily and independently scaled. The deduced preliminary mass-resolution amounts to about 5 a.m.u (FWHM).

axis of the chamber. The distance from the symmetry axis of the chamber to the entrance window of the HPGe detectors amounts to 11 cm. The gamma detection efficiency has been determined using standard gamma sources, as well as MCNP [4] simulations accounting for gamma attenuation in the ionization chamber material. The total gamma detection efficiency is maximum at about 150 keV and amounts to 6.2%. It amounts to 1.5% at 1300 keV.
The trigger of the data acquisition system consists of a valid fission event for which both anodes of the ionization chamber detect a signal above the electronic threshold. The later is set just above the ^{252}Cf alpha particles signal. For each valid event, the kinetic energies of both fission fragments are recorded. When present the energy signal of the HPGe detectors are also recorded as well as the time elapsed between the fission event and the detection of the corresponding gamma-ray.

The measured kinetic energies are calibrated in such a way that the mean energy of the light and heavy fragment peaks correspond to the literature values. The present results are still preliminary and corrections for energy-loss in the ^{252}Cf sample; pulse height defect resulting from non-ionizing collisions in the gas and for neutron emission are not yet included. For more details on the complete analysis procedure, the interested reader can refer to Ref. [3]. Figure 2 presents the preliminary post-neutron mass-distribution of the fission fragments deduced from the measured kinetic energies. The blue curve on the figure corresponds to the same mass-distribution gated by the detection of a gamma-ray belonging to the decay of the 6^+ isomer in ^{134}Te. From the later distribution one deduces a preliminary mass-resolution of about 5 a.m.u. (FWHM) that, combined with the standard energy resolution of the HPGe detectors, provides the selectivity of the present experimental setup. As an example, figure 3 (a)

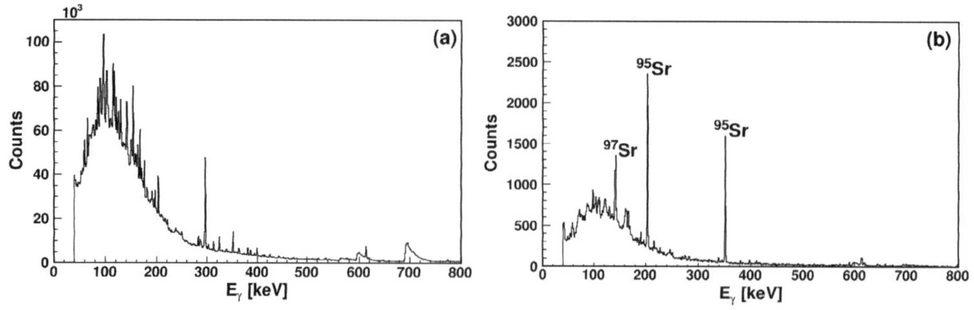

Figure 3. (a) Single delayed gamma spectrum measured in coincidence with a fission event in the ionization chamber. The gamma line is detected from 10 up to 200 ns after fission. (b) Same as (a) with the additional condition that the mass of the light fission fragment is A = 95 ±0.5.

shows the single delayed gamma spectrum measured from 10 ns up to 200 ns after a fission event, while figure 3 (b) shows the same spectrum gated by mass A = 95 in the ionization chamber. In panel (b) one clearly sees the 204 and 352 keV prominent peaks belonging to the decay of the $7/2^+$ isomer in ^{95}Sr. These gamma-lines are hardly visible in panel (a) demonstrating the selectivity of the mass gate. Because of the aforementioned 5 a.m.u. mass resolution, one also observes a gamma line belonging to mass A = 97 in the spectrum of figure 3 (b).

3 Preliminary results

In this section we present the procedure used in order to extract isomeric production yields. For each isomer, we construct the mass-distribution of its characteristic gamma-lines. As an example the mass distribution for the 1610 keV isomer in ^{133}Te is reported in figure 4. Each point of the mass distribution is obtained by extracting the number of counts in the characteristic peaks of the mass-gated gamma spectrum. In order to maximize the peak to background ratio, gamma spectra are restricted to events detected up to 3 to 5 times the lifetime of the considered isomer. The resulting experimental mass distribution is then adjusted with a Gaussian function in order to extract the total number of counts in the considered gamma peak. This total number of counts is corrected for the aforementioned time gate applied to the gamma spectrum as well as for the gamma detection efficiency. The analysis is performed for each side of the twin ionization chamber, both sides yielding compatible number of counts within the statistical error bars. The production yield of the isomer is obtained by normalizing the total number of counts in the gamma line of interest to the number of fission events detected in the ionization chamber, considering that each fission event gives rise to two fission fragments.

When applied to the 1610 keV isomer in ^{133}Te, this procedure results in a preliminary production yield of $Y = 1.23\times10^{-2}(5)$, to be compared to the total production yield of $Y = 1.824\times10^{-2}$ for ^{133}Te according to the Wahl evaluation [5]. In the future this value will be compared to theoretical ones obtained, for example, with codes such as FIFRELIN [6] or CGMF [7, 8] aiming at describing the decay of fission fragments. The tentative spin/parity of the 100 ns isomer discussed above is $19/2^-$ while the ground state spin/parity of the nucleus is $3/2^+$. From the large spin difference between the isomer and the ground state one expects that their relative feeding in the fission process somewhat reflects the characteristics

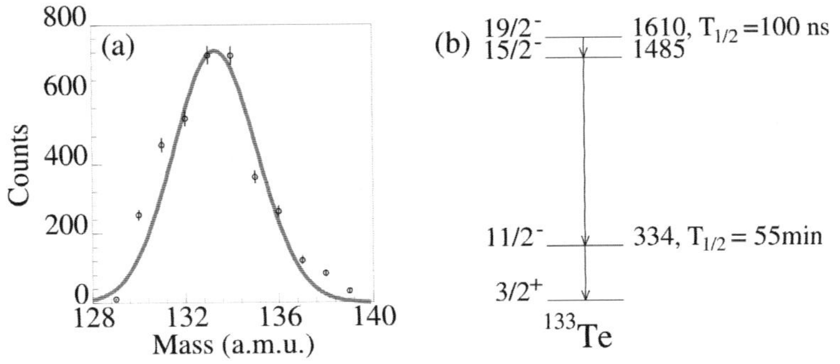

Figure 4. (a) Mass distribution associated with the 1150 keV γ transition belonging to the decay cascade of the 100 ns isomer lying at 1610 keV in ^{133}Te. The curve is fitted against the experimental data points. (b) Partial level scheme of ^{133}Te showing the decay path of the 100 ns isomer at 1610 keV.

of the spin distribution in the primary fission fragments and that present data could constraint related model parameters.

4 Conclusion

In this contribution we report on preliminary results on production yields of micro-second isomers populated in ^{252}Cf(SF). The experimental setup consisting of a twin Frisch grid ionization chamber surrounded by 8 HPGe detectors has been presented. The selectivity of the setup for gamma-ray spectroscopic purposes has been demonstrated. About 60 isomeric states have been observed with this setup. The procedure developed in order to extract the production yields of these isomers has been presented for the particular case of ^{133}Te.
In the future this procedure will be applied to all the isomers observed in this work. The deduced production yields will be compared to theoretical ones in order to provide constraints on parameters for modeling spin distribution in primary fission fragments.

References

[1] L. Gaudefroy *et al.*, Phys. Rev. C **97**, 064317 (2018)
[2] S. Pauker and N. H. Steiger-shafrir, Nucl. Instrum. Methods A **91**, 557 (1971)
[3] L. Gaudefroy *et al.*, Nucl. Instrum. Methods A **855**, 133 (2017)
[4] C. J. Werner *et al.*, "MCNP6.2 Release Notes", Los Alamos National Laboratory, report LA-UR-18-20808 (2018)
[5] A. C. Wahl, At. Data Nucl. Data Tables **39**, 1 (1988)
[6] O. Litaize, O. Serot and L. Berge, Eur. Phys. J. A **51**, 177 (2015)
[7] T. Kawano, P. Talou, M. B. Chadwick and T. Watanabe, J. Nucl. Sci. Technol. **47**, 462 (2010)
[8] P. Talou, B. Becker, T. Kawano, M. B. Chadwick and Y. Danon, Phys. Rev. C **83**, 064612 (2011)

https://doi.org/10.1051/epjconf/202024201004

Measurements of short-lived fission product yields using monoenergetic neutron and photon beams

Sean Finch[1,2,3,*], *Matthew* Gooden[4], *Chris* Hagmann[3], *Calvin* Howell[1,2], *Vanessa* Linero[4,5], *Jack* Silano[3], *Mark* Stoyer[3], *Anton* Tonchev[1,3], *Werner* Tornow[1,2], *Innocent* Tsorxe[2,6], and *Jerry* Wilhelmy[4]

[1]Duke University, Durham, NC 27708
[2]Triangle Universities Nuclear Laboratory, Durham, NC 27708
[3]Lawrence Livermore National Laboratory, Livermore, CA 94550
[4]Los Alamos National Laboratory, Los Alamos, NM 87544
[5]Colorado School of Mines, Golden, CO 80401
[6]North Carolina State University, Raleigh, NC 27695

Abstract. A joint TUNL-LLNL-LANL collaboration was formed to measure the absolute fission product yields from the ^{235}U, ^{238}U, and ^{239}Pu isotopes. Our goal is to study the energy evolution of fission products using monoenergetic beams. In order to extend our successful fission product-yield studies to include products with shorter half-lives, a RApid Belt-driven Irradiated Target Transfer System, named RABITTS, was constructed. This system allows us to perform cyclic activation and quantify fission products with γ-ray spectroscopy using HPGe detectors. Both a 1 meter and 10 meter transfer system have been developed, with transit times of 0.4 and 1.1 seconds, respectively. Using these systems, we have measured sub-second half-lives. Our goal is to measure fission product yields from neutron-induced fission with E_n = 0.5 – 14.8 MeV and photofission with E_γ = 8 – 15 MeV. A detailed characterization of the system's performance is presented, including preliminary fission product measurements, and the expected sensitivity.

1 Introduction

Fission product yields (FPYs) are a principal observable of the fission process. The dominant nuclear fission process produces two fission fragments, which promptly evaporate neutrons and result in two fission products. These fission products are created far from nuclear stability, and usually undergo multiple decays before reaching stability. The FPY is the percentage production of a given fission product. Independent FPYs are the initial product yields following post scission neutron emission. The cumulative FPY of a particular isotope is the sum of its independent yield plus the independent yields of all unstable precursor isotopes that β decay into that isotope, as shown in Fig. 1.

An experimental advantage of FPYs is that they provide fundamental fission information that can be obtained using radiochemical or counting techniques at a time that is delayed relative to the prompt fission event. A joint TUNL-LLNL-LANL collaboration has produced

*e-mail: sfinch@tunl.duke.edu

such measurements of absolute FPYs [1, 2], primarily focusing on cumulative yields. The experimental goal of this program is to measure the energy dependence of FPYs as a function of incident neutron energies from $E_n = 0.5$ to 14.8 MeV using monoenergetic beams [2]. Most FPY evaluations only consider three incident neutron energies: thermal, fission spectrum ($\approx$ 1.5 MeV), and 14.8 MeV [3]. We are also pursuing additional measurements to include photofission product yields at $E_\gamma = 8 - 15$ MeV [4]. The neutron induced fission measurements are performed on the main three actinides: ^{235}U, ^{238}U, and ^{239}Pu, using the same facilities and techniques. Great detail is taken to perform these experiments in such a way as to minimize systematic uncertainties between the different measurements. These results are an extensive self-consistent nuclear data set of FPYs.

The goal of the present work is to extend these cumulative FPY measurements towards independent FPY measurements. This involves measuring fission products with progressively shorter half-lives. The technique to be used for these measurements is detailed in Sec. 1.1. and preliminary fission spectra are presented in Sec. 2.1.

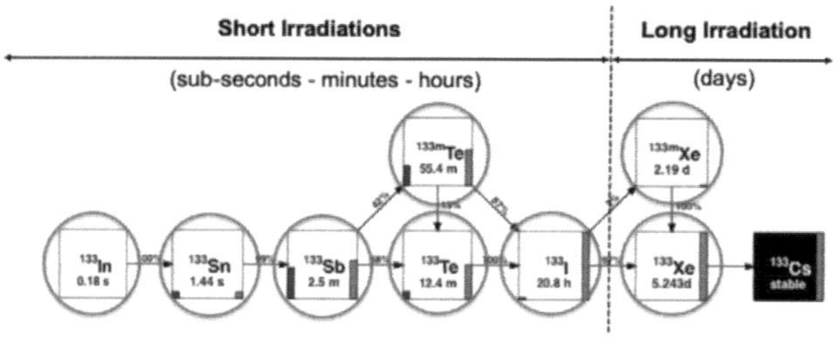

Figure 1. The A = 133 isobar, illustrating the possible fission products. The cumulative FPY of the entire $A = 133$ isobaric chain is measured using the yield of ^{133}Xe. Additional measurements of the shorter-lived fission products will be able to provide information on the independent yields.

1.1 Experimental technique

Quasi-monoenergetic neutron beams were produced at Triangle Universities Nuclear Laboratory's (TUNL) 10 MV tandem accelerator using the ^{7}Li$(p,n)^7$Be, ^{3}H$(p,n)^3$He, ^{2}H$(d,n)^3$He, and ^{3}H$(d,n)^4$He reactions. For the photofission studies, the High Intensity γ-ray Source (HIγS) [5] was used. HIγS produces monoenergetic γ rays in the energy range from 2 to 100 MeV from the Compton back-scattering of free electron laser light in an electron storage ring. The typical γ-ray flux delivered on targets for this work was 4×10^8 γ/(cm^2 s).

FPYs are determined via the activation technique: following irradiations, the targets are γ-ray counted using high purity germanium (HPGe) detectors. The fission products are tracked over multiple half-lives. This allows for precise measurement of the fission product's characteristic γ-ray energy spectra and half-life, which helps to unambiguously identify the exact fission product. The total number of fissions in the activation target is determined using a dual fission chamber, and mass scaling the thin fission chamber foils to the thick activation target. The power of this technique is that it does not rely on measurement of the exact beam flux or any reference cross-section data. The technique is characterized in Ref. [6], and updated Monte-Carlo simulations of the target geometry have been developed for the present work.

The original measurements focused primarily on cumulative FPYs [1, 2, 4]. These measurements, as shown in the right side of Fig. 1, typically used 3-5 day long irradiations, followed by 1-2 month counting cycles, and resulted in measurements of 16 FPYs with

$t_{1/2} > 0.5$ h. Shorter 1 h long irradiations were then performed to measure yields of an additional 45 fission products with $t_{1/2} > 3$ min. The minimum observable $t_{1/2}$ is limited by the transit time from the activation source to the low background counting stations located in an adjacent building. In order to reduce the delay time between the end of target irradiation and the beginning of counting to the shortest possible value, a new experimental capability has been added to TUNL: a RApid Belt-driven Irradiated Target Transfer System (RABITTS).

1.2 RABITTS

RABITTS is a fully automatic system that moves a target between irradiation and counting positions in order to perform cyclic activation [7, 8]. The user may program irradiation times, counting times, and transit speeds. FPYs have previously been measured using cyclic activation and a fast transit system at $E_n = 14.8$ MeV [9]. A total of three RABITTS were constructed: a 1 m and a 10 m track for neutron-induced fission, and a 1 m track for photofission studies. These three systems are identical except for track length and installation location. RABITTS has a maximum speed of 10 m/s, which results in a 0.4 s transit time for the 1 meter systems, and a 1.1 s transit time for the 10 m system.

The RABITTS is powered by a servo motor attached to a belt-driven linear actuator track. The servo motor and linear actuator has a position repeatability of 34 μm and 25 μm, respectively. This combined system results in a 42 μm repeatability for our target sample. Cyclic activation has traditionally been performed using pneumatic systems. The present approach, using a servo motor, track, and belt, was chosen for its high repeatability, controlled soft acceleration and deceleration. Furthermore, there are no restrictions on the target mass, size, or shape, and the target remains in a consistent orientation throughout the cycle. For the present experiments, thin targets with a total mass of 400-500 mg have been utilized. The use of these relatively thin targets minimizes corrections such as multiple scatter effects and self γ-ray attenuation.

A schematic of the 10 m RABITTS is shown in Fig. 2. The actinide target is irradiated in the center of a large 10 m × 10 m room, then transported through a thick concrete shielding wall to a low background counting area. In the present work, two HPGe detectors are used, with the capability to include two additional HPGe detectors during future experiments.

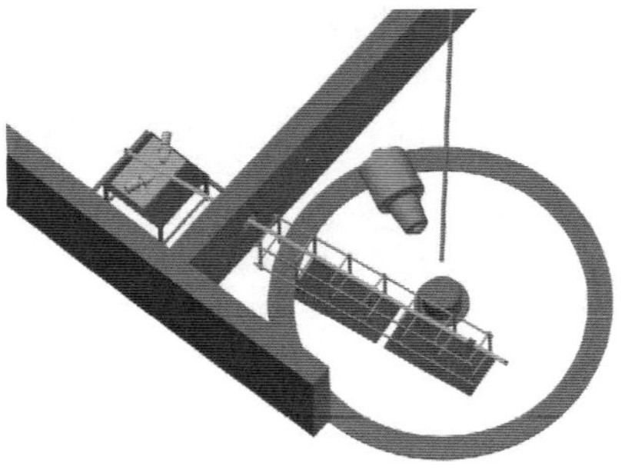

Figure 2. A schematic of the 10 m RABITTS. The track extends from the counting station with two HPGe detectors on the left side, through the concrete shielding wall, to the irradiation position on the right side.

A new digital data acquisition (DAQ) system was utilized for this project. All signals are recorded in list-mode electronics and time stamped with sub-microsecond precision. This includes the HPGe detectors, fission chambers, servo motor movements, a liquid scintillator neutron flux monitor, and a beam current indicator. The system allows for $\gamma - \gamma$ coincidence analysis between the HPGe detectors. The servo motor controller is programmed to output a signal when the target leaves and arrives at each position. Thus every event can be precisely tagged in the activation cycle.

The RABITTS is also used to pulse the incident neutron beam. The servo motor is programmed to control the voltage supply of a steerer magnet located in the transport section of the charged particle beam (p or d). This steerer deflects the beam so no neutrons are produced during the counting and transit periods. This procedure provides a sharp cut off for the irradiations, ensures no additional irradiation occurs in transit, and reduces any neutron-induced backgrounds observed in the HPGe detectors. The neutron flux is measured using a liquid scintillator neutron detector downstream of the target. During counting and transit cycles, this detector measures background rates. After the target returns to the irradiation position, the neutron flux will return to is maximum value over 50 ms. Although this time is very fast compared to our measured $t_{1/2}$, the beam rise time is accounted for as a beam fluctuation correction in our analysis procedure. This system is fully synchronized by design, and is not sensitive to drifts between the DAQ, servomotor, or beam pulsing system. The DAQ system was used to measure the total cycle time over 1000 cycles, and determined a total deviation of <0.2 ms. Nonetheless, we measure and correct for this small timing deviation during analysis.

2 Testing the system

Prior to measurements of fission spectra, an extensive campaign was undertaken to benchmark and characterize the RABITTS. This ensures our FPY values will be of highest quality and fidelity. The system was calibrated using standard γ-ray calibration sources, such at ^{137}Cs. To test the performance of the system with beam, we measured neutron-induced isomer production. The $t_{1/2}$ = 7.75 s ^{197m}Au isomer was first measured. The strong characteristic γ-ray transitions from this source are clearly seen (Fig. 3), following irradiation with E_n = 2.0 MeV neutrons. The preliminary measured $t_{1/2}$ = 7.73 ± 0.06 s agrees with the literature value [10].

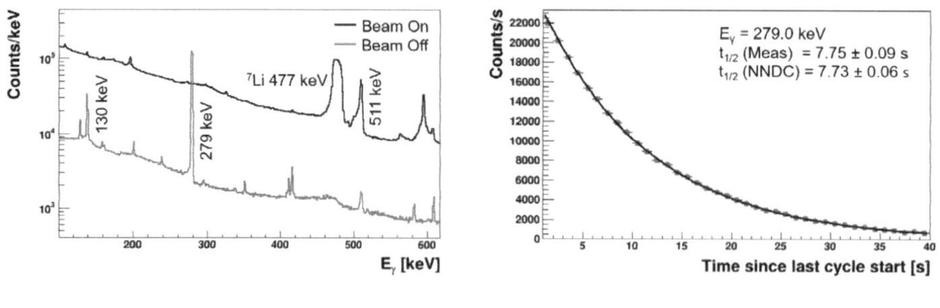

Figure 3. Measurement of the ^{197}Au$(n, n')^{197m}$Au reaction. The measured γ-ray spectra is shown with the characteristic peaks at 130.2 and 297.01 keV. Note the reduction in background between the irradiation and counting cycles due to the beam pulsing system. The decay curve of ^{197m}Au is also shown, with a preliminary $t_{1/2}$ = 7.73 ± 0.06 s

These data were taken with the 1 m RABITTS. Although the HPGe detectors are well shielded, they are sensitive to the large neutron flux, as seen in the beam on and beam off spectra of Fig. 3. The effectiveness of the beam pulsing system may be seen between the two spectra. By moving to the 10 m RABITTS, the issue is further alleviated, reducing any residual activation in the HPGe detector components.

To test the lower limits of the RABITTS, the $t_{1/2}$ = 809 ms ^{90m}Zr isomer was also studied. Here the isomer life time closely matches the transit time, providing a good sensitivity measurement. Performing cyclic activation with the 1 m RABITTS at E_n = 5.0 MeV produced the γ-ray spectra and decay curve shown in Fig. 4. The preliminary measured $t_{1/2} = 787 \pm 8$ ms is notably shorter than the literature value of $t_{1/2} = 809 \pm 2$ ms [11].

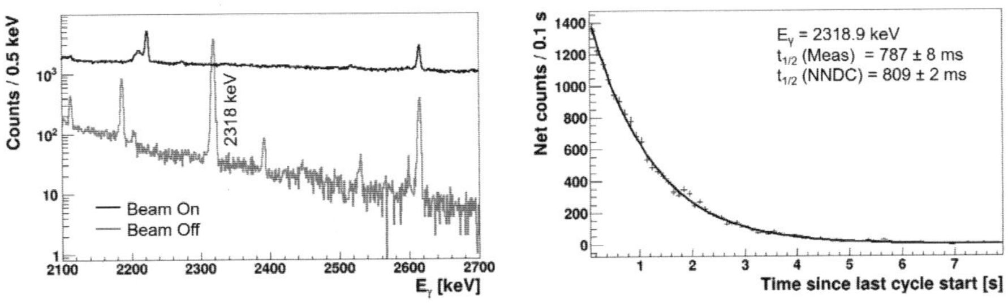

Figure 4. Measured γ-ray spectra showing the 2186.2 and 2318.9 keV transitions from the decay of ^{90m}Zr, and the resulting decay curve of the 2318.9 keV γ-ray yield.

2.1 First fission data

Immediately following the previous isomer production tests, we proceeded with measurements of fission spectra using RABITTS. Focusing on short-lived fission products, strong γ rays from the decay of ^{96}Sr ($t_{1/2}$ = 1.07 s), ^{99}Zr ($t_{1/2}$ = 2.1 s), and ^{97m}Y ($t_{1/2}$ = 1.17 s) were immediately observed. The ^{97m}Y decay is shown in Fig. 5 using the 10 m RABITTS with a 4 s irradiation, 1.1 s transit, and 8 s counting interval. The short $t_{1/2}$ = 1.17 s fission product is still strongly observed, even given the relatively long 1.1 s transit time associated with the 10 m track. Cycle times were adjusted to investigate fission products with $t_{1/2}$ from 1 to 300 s. During the FPY measurements, cyclic activation of actinide targets were conducted during the day, and an > 8 h counting cycle was performed overnight. This overnight counting period helps to reduce activity from long-lived fission product build up in the target. Furthermore the decays from the long-lived fission products that have been produced in the irradiation are also measured by the HPGe detectors, resulting in measured yields for products with $t_{1/2}$ < 8 h. This permits us to seamlessly connect very short (independent) yields with long-lived (cumulative) yields in the same experiment.

3 Conclusions

Over 300 γ-ray lines, representing over 50 unique fission products, have been observed by performing cyclic activation with RABITTS. Data has been taken on ^{235,238}U, ^{239}Pu for neutron-induced fission at E_n = 2.0, 4.5 MeV, and photofission at E_γ = 11.2, 13.0 MeV. Analysis on these data sets is currently underway. Future plans include the incorporation of the fission induced electromagnetic response (FIER) [12] code to investigate fission products

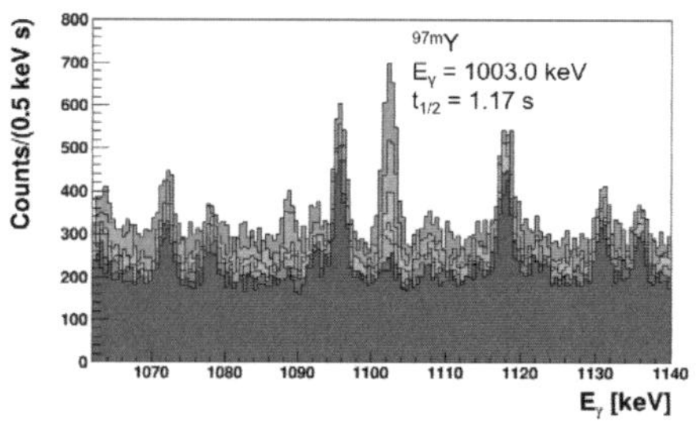

Figure 5. A measured γ-ray spectra taken using cyclic activation of $^{238}U(n, f)$ at E_n = 2.0 MeV, with 4 s irradiation, 1.1 s transit, and 8 s counting time. Each histogram represents a 1 s counting interval. The strong 1003.0 keV line results from the decay of ^{97m}Y + ^{97}Y, with half-lives of 1.17 and 3.75 s, respectively.

and extract yields. We have demonstrated the ability to measure yields of fission products with $t_{1/2}$ from 1 s to 300 s. These data will connect to our previously measured cumulative FPYs of $t_{1/2} > 300$ s to produce a complete, self-consistent nuclear data set. RABITTS is a new capability added to the experimental facilities at TUNL. In addition to its primary objective of measuring FPYs, RABITTS is a robust system for measuring neutron or γ-induced nuclear reactions that product radioactive products. It allows us to compare FPY data from two different nuclear reactions, where both mechanisms are studied under the same experimental conditions. It will extend our fundamental knowledge of nuclear fission by providing highly accurate FPY data for theoretical models using two electrically neutral incident particles: the neutron and photon, which induce fission via the strong nuclear interaction and by electromagnetic stimulation, respectively.

4 Acknowledgments

This work was supported in part by the National Nuclear Security Administration Stewardship Science Academic Alliances grant no. DE-NA0003884, DE-NA0003887, and the U.S. Department of Energy, Office of Nuclear Physics, under grant no. DE-FG02-97ER41033. The work was performed under the auspices of the US Department of Energy by Los Alamos National Laboratory and Lawrence Livermore National Laboratory under Contract No. DE-AC52-07NA27344. We would like to thank Todd Bredeweg and Krishichayan for their assistance to this project.

References

[1] C. Bhatia et al., Phys. Rev. C 91, **064604**, (2015)
[2] M. Gooden et al., Nucl. Data Sheets **131**, 319 (2016)
[3] T.R. England and B.F. Rider, "Evaluation and compilation of fission product yields," ENDF-349, LA-UR-94-3106 (1994)
[4] Krishichayan et al., Phys. Rev. C 100, **014608**, (2019)
[5] H.R. Weller et al., Prog. Part. Nucl. Phys. **62**, 275 (2009)
[6] C. Bhatia et al., Nucl. Instr. Meth. Phys. Res. A **757**, 7 (2014)
[7] W.W. Givens, W.R. Mills, and R.L. Caldwell. Nucl. Instr. Meth. **80**, 95 (1970)
[8] N.M. Spyrou. J. Radioanal. Chem. **61**, 211 (1981)

[9] B. Pierson et al., Nucl. Data Sheets **139**, 171 (2017)
[10] H. Xiaolong and Z. Chunmei, Nucl. Data Sheets **104**, 283 (2005)
[11] E. Browne, Nucl. Data Sheets **82**, 379 (1997)
[12] E.F. Matthews et al, Nuclear Inst. and Methods in Physics Research, A **891**, 111 (2018)

Fusion-induced fission measurements with the MUSIC active target detector

Sergio Almaraz-Calderon[1,*], *K. Ernst* Rehm[2], *Benjamin W.* Asher[1], *Melina L.* Avila[2], *Kalle* Auranen[2], *Birger B.* Back[2], *Jie* Chen[2], *Clayton* Dickerson[2], *Cheng-Lie* Jiang[2], *Richard C.* Pardo[2], *Daniel* Santiago-Gonzalez[2], *Guy* Savard[2], and *Rashi* Talwar[2]

[1]Department of Physics, Florida State University, Tallahassee, FL 32306, USA
[2]Physics Division, Argonne National Laboratory, Argonne, IL 60439, USA

Abstract. The rapid neutron capture process (*r-process*) is believed to be responsible for about half of the production of the elements heavier than iron and it may also contribute to abundances of some lighter nuclides. Great excitement was recently generated by evidence for *r-process* nucleosynthesis in binary neutron star mergers via multi-wavelength observations of kilonova emission and gravitational waves. In order to interpret the observations and validate theoretical predictions, an understanding of the fission process, in particular of the evolution of fission barrier heights, is needed. An experimental study of fusion-induced fission cross sections using active-target detectors is a promising idea since the fission excitation function can be studied with a single beam energy. The Multi-Sampling Ionization Chamber (MUSIC) is an active-target detector in which a gas serves as both, counting gas and target nuclei. A proof-of-principle experiment to explore the ability to identify fission events with MUSIC was recently performed at Argonne National Laboratory. In this work, ideas, results and perspectives will be discussed.

1 Introduction

Understanding of the fission process is crucial in applied as well as fundamental research. Fission models are needed for nuclear energy applications, nuclear structure and nuclear astrophysics studies. Theoretical predictions of the fission barriers of exotic systems drive the quest for super-heavy nuclei [1]. The reliable determination and theoretical prediction of fission barriers, fission fragment distributions, as well as the balance between the neutron separation energy and fission-barrier heights, are decisive for determining the behavior of different isotopes in a nuclear reactor as well as the nature of the termination of stellar nucleosynthesis in the *r-process* [1, 2]. The *r-process* path ends when it reaches an area in the nuclear chart where fission dominates over further neutron captures. The position of this point is determined by the fission barriers and the neutron separation energies of the nuclei involved. The *r-process* reaction path proceeds through very neutron-rich nuclei that cannot (yet) be studied experimentally. This means that *r-process* reaction networks need to rely on theoretically predicted nuclear properties [3, 4]. The fission probability of a nucleus depends

*e-mail: salmarazcalderon@fsu.edu

directly on the fission barrier and, thus, fission barriers are crucial to determine the end of the *r-process* path, having a direct influence on the final abundance distribution.

Experimental investigations are needed to validate the theoretical efforts and provide benchmark information to reliably extrapolate theoretical calculations, particularly around closed shells where barrier heights are expected to rise, greatly influencing the path of the *r-process* [4]. Fission cross section measurements using active-target detectors are a promising idea. The measurements can be performed with a single beam energy by studying the fission excitation function as the beam travels through the detector. Recently, a proposal has been accepted by the NSCL PAC to measure fission using the AT-TPC at NSCL [5]. A fission TPC is also being developed by the NIFFTE collaboration at LANL [6]. A measurement using the MAYA active-target detector of fission induced by the ^{238}U+^{12}C system was recently performed at GANIL [2].

We are currently investigating the ability to measure fusion-induced fission excitation functions with the MUSIC detector.

2 The MUlti Sampling Ionization Chamber (MUSIC) detector

The MUlti Sampling Ionization Chamber (MUSIC) detector [7] is an active target detector developed at Argonne National Laboratory (ANL) that measures energy losses as the beam passes through the detector. A segmented anode allows for a measurement of an excitation function covering a large energy range using a single beam energy. The range of the excitation function is determined by the beam energy and the pressure of the gas inside the detector. The anode is segmented in 18 strips. The first and last strips are used as control strips. The other 16 strips are subdivided in left and right to measure the multiplicity of events. MUSIC was originally designed to measure fusion events. A sudden increase in the energy loss in the strip where the reaction occurs is a signature of a fusion event. MUSIC can be filled with several gases e.g. methane, helium, neon and argon. MUSIC has been successfully used to measure fusion excitation functions of exotic beams [8] as well as (α,p) and (α,n) reactions [9, 10].

3 Proof-of-Principle Experiment

A proof-of-principle experiment to explore the ability to identify fission events in MUSIC was already performed at the ATLAS facility at ANL. A 1000 MeV ^{165}Ho beam was used to bombard MUSIC which was filled with ^{40}Ar gas at 113 Torr. In MUSIC, a fusion-induced fission reaction is identified by a sudden 'jump' in the energy loss signal characteristic of fusion, followed by two big signals in both sides of the detector from the two fission fragments. A schematics of fusion-fission events along with the schematics of the segmented anode in MUSIC are shown in figure 1.

Fission fragments induced by the ^{165} Ho+ ^{40} Ar$\rightarrow$ ^{205} At reaction in the $E_{cm} = 130-160$ MeV energy range in the center of mass were identified by their traces in MUSIC. A simulation of the traces from fission reactions occurring in MUSIC was performed and compared with the experimentally extracted traces showing a good agreement between them. The preliminary experimental cross sections of the fusion-induced fission from the ^{165} Ho+ ^{40} Ar$\rightarrow$ ^{205} At reaction measured with MUSIC are shown in figure 2. Previously measured data is also shown for comparison [11]. The good agreement give us confidence in the ability to reliably identify fission events in MUSIC. Fission cross sections were also calculated with the code PACE4 [12]. The experimental values for the fission cross sections measured with MUSIC and previously reported data from Ref. [11] are within 20% of the fission cross sections calculated with PACE4. However, a significant disagreement was observed

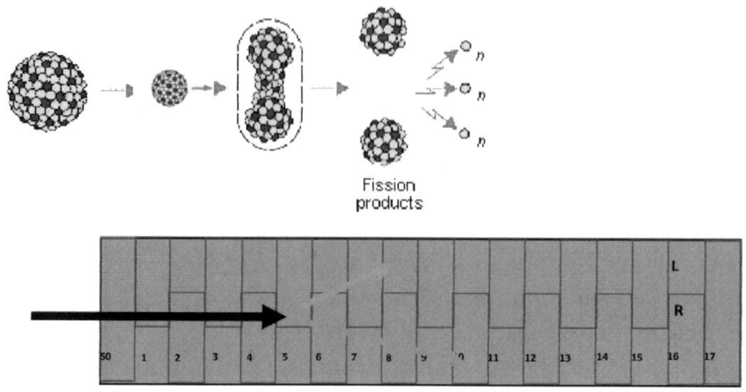

Figure 1. Schematics of a fusion-induced fission event and the segmented anode in MUSIC representing a fission event occuring in strip 5.

when comparing experimental data from the literature with fission cross sections calculated with the PACE4 for other systems. The PACE4 calculated cross sections were zero despite existing experimental data with significant cross sections as shown in table 1. Default settings were used for such calculations, same that predicted well the cross section for the ^{205}At case. It is therefore crucial to validate and benchmark the theoretical predictions for fission cross sections where experimental data is available in order to reliably extrapolate to exotic regions where experiments are not yet possible.

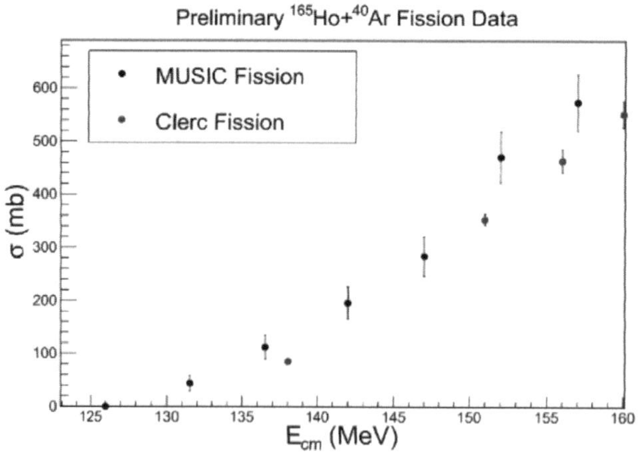

Figure 2. Preliminary experimental cross section of fusion-induced fission events measured in MUSIC for the ^{165}Ho+^{40}Ar→ ^{205}At system compared with data from Ref. [11].

Table 1. Experimental fission cross sections reported in the literature compared with PACE calculations for different systems. A strong disagreement is observed for some systems.

System	E_{cm} (MeV)	Experimental σ (mb)	PACE Calculated σ (mb)
^{165}H0+^{40}Ar→ ^{205}At	156	464 ± 22 [11]	361
	160	552 ± 25 [11]	416
	166	596 ± 29 [11]	510
^{96}Mo+^{40}Ar→ ^{136}Nd	140	141 ± 30 [14]	0
	209.5	160 ± 15 [14]	0
^{110}Pd+^{40}Ar→ ^{150}Gd	144	290 ± 40 [13]	0
	165	380 ± 40 [13]	0
	192	610 ± 60 [13]	0

4 Summary and Outlook

Fusion-induced fission reactions can be studied with the MUSIC active target detector. For the near future, stable Mo beams from ATLAS and radioactive ^{104,106}Mo beams from CARIBU would be used to bombard the MUSIC detector filled with Argon gas. A systematic study the fission excitation functions and the evolution of the fission barrier around neutron number $N \sim 82$ would be performed. The rates of the ^{104,106}Mo CARIBU beams (a few thousand pps) are ideal for this study. A comparison with theoretical predictions would allow to benchmark fission models. This is the first step towards a systematic study of the evolution of fission barriers around neutron closed shell $N = 128$, where the strongest influence on the *r-process* path is expected [1, 4]. Future measurements with other CARIBU beams will be proposed in preparation for beams from FRIB.

Acknowledgements

This work was partially supported by the U.S. Department of Energy, Office of Nuclear Physics, under contract No. DEAC02-06CH11357, the National Science Foundation under grant No. PHY-1712953, the Stewardship Science Academic Alliance through the CENTAUR Center of Excellence under grant no. DE-NA0003841. This research used resources of ANL's ATLAS facility, which is a DOE Office of Science User Facility.

References

[1] P. Möller et al., Phys. Rev. C **91**, 024310 (2015).
[2] C. Rodriguez-Tajez et al., Nucl. Phys. A **958**, 246 (2017).
[3] P. Möller et al., Phys. Rev. C **79**, 064304 (2009).
[4] N. Vassh et al., J. Phys. G: Nucl. Part. Phys. **46**, 065202 (2019).
[5] NSCL-PAC, https://publicapps.nscl.msu.edu/completedExperiments/experiments/view/672.
[6] M. Heffner et al., Nucl. Instrum. Meth. Phys. Res. A **759**, 50 (2014).
[7] P. F. F. Carnelli et al., Nucl. Instrum. Meth. Phys. Res. A **799**, 197 (2015).
[8] P. F. F. Carnelli, et al., Phys Rev Lett. **112**, 192701 (2014).
[9] M. L. Avila et al., Phys. Rev. C **94**, 065804 (2016).
[10] M. L. Avila et al., Nucl. Instrum. Meth. Phys. Res. A **859**, 63 (2017).
[11] H.-G. Clerc et al., Nucl. Phys. A **419**, 571 (1984).
[12] A. Gavron, Phys.Rev. C **21**, 230 (1980).

[13] C. Cabot et at., Nucl. Phys. A **427**, 173 (1984).
[14] B. Tamain et al., Nucl. Phys. A **252**, 187 (1975).

Bragg curve spectroscopy for improved fission fragment identification

Bryan Fulsom[1,*]

[1]Pacific Northwest National Laboratory, Richland WA 99352, USA

Abstract. We report on the development of Bragg curve spectroscopy techniques to improve fission fragment identification in the measurement of independent fission product yields. The NIFFTE collaboration's fissionTPC detector provides ionization energy and particle tracking information from neutron-induced fission targets. A joint effort between PNNL, LLNL, LANL, and the Colorado School of Mines is investigating the ionization profiles deposited by U-235, U-238, and Pu-239 fission products in this detector, with the goal of including additional stopping power information beyond a standard 2E analysis. The aim is to improve the determination of fragment atomic and mass numbers with this information, via methods such as parametric fits and machine learning techniques.

1 Introduction

This talk focuses on fission product yield measurements made with the NIFFTE TPC detector. It first covers a description of the detector, then provides detailed information on the application of the 2E method to measure fission product yields from a ^{235}U target, with preliminary results. We then discuss the use of dE/dx information afforded by the tracking capabilities of the NIFFTE TPC to improve fission product identification. Finally, the initial application of machine learning techniques to improve this effort is presented.

2 The NIFFTE TPC

The Neutron Induced Fission Fragment Tracking Experiment (NIFFTE) is a collaboration between seven U.S. institutions to construct the "fissionTPC" time projection chamber located at the Los Alamos Neutron Science Center (LANSCE). The detector was designed to perform precision measurements of fission cross section ratios using tracking capability to reduce systematic uncertainties. For a sample of recent NIFFTE collaboration publications see [1–3].

The fissionTPC is a dual-volume gas time projection chamber (see Fig. 1). The anode detection planes consist of 2976 hex pad MicroMEGAS detectors with 2mm pitch. The ionization gas volume is filled with a 95% argon / 5% isobutane mix, operated at 560 torr. The drift distance in the direction parallel to the neutron beam is 54mm. Signals are readout with a data acquisition system with 50 MHz pad readout, and 1 GHz for the cathode used

*e-mail: bryan.fulsom@pnnl.gov

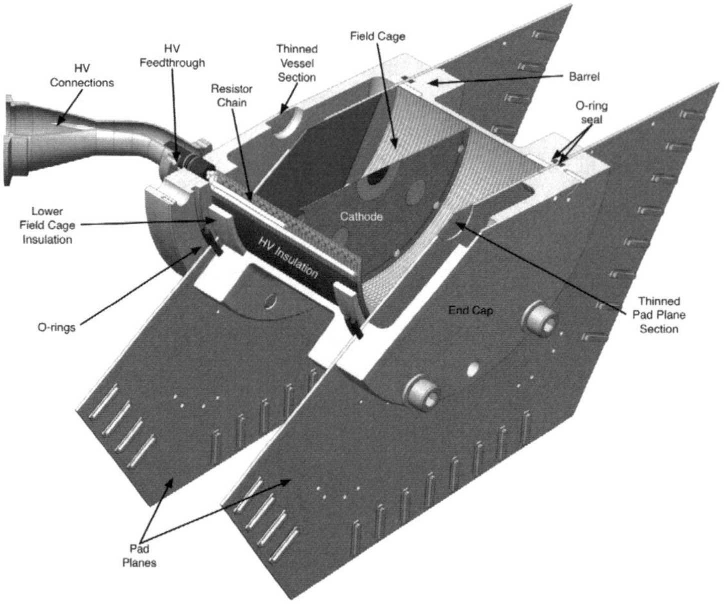

Figure 1. Schematic layout of the NIFFTE fissionTPC time projection chamber.

for neutron time-of-flight (energy) measurements. Publications are available detailing the NIFFTE detector [4], data acquisition [5], and operation [6].

The NIFFTE TPC is situated on the LANSCE Weapons Neutron Research facility 90L beamline, which provides an unmoderated source of neutrons ranging in energy from ~0.1 to 100 MeV (though in practice, given the flux and region of interest, energies up to $O(10)$ MeV are considered). Here we consider data collected from the neutron beam on targets of ^{235}U, ^{238}U, and ^{239}Pu, taken during 2016 and 2017. The 100μg/cm^2 targets were prepared at Oregon State University by vapor deposition on a thin (~100μg/cm^2) carbon backing.

Time projection chambers provide the ability to fully reconstruct the three-dimensional path of particle tracks. From these tracks, one can measure the direction, energy, starting vertex position, and length. The energy loss per unit distance along the track (Bragg curve), dE/dx, can also be measured. The combination of these properties allows the NIFFTE TPC measurements to precisely select fission fragments eminating from mixed isotope targets, discriminate between proton, α particles, and fission fragments based on track length and energy, and potentially mass (A) and nuclear charge (Z) from details of the Bragg curve.

3 Application of the 2E Method to NIFFTE TPC data

This section provides details of the application of the "2E Method" to NIFFTE TPC data.

Events are considered for further study if there is at least one track reconstructed in both volumes, with a minimum ADC count (energy) of 3500, originating from the ^{235}U target location. The tracks must fall within $0.30 < \cos\theta < 0.92$ ($23^\circ < \theta < 73^\circ$), where θ is the angle between the track path and the neutron beam axis. This selection removes fragments that pass through large amounts of target/backing material (steep θ) and those whose signals become saturated in the DAQ system (shallow θ). Finally, based on studies of anode pad efficiency, tracks entering regions of "dead" and low-efficiency pads are removed, in order to maintain the highest quality data sample.

Energy loss in the target and backing is the largest source of correction and systematic uncertainty in this analysis. A correction is applied according to the angle of the track, as a proxy for the amount of thickness of material traversed. The energy loss correction is estimated from a linear fit to the number of ADC counts versus $1/\cos\theta$ for all tracks in the data set. Separate corrections are applied for each volume, and can range between 10-30 MeV for fission fragments passing through the target carbon backing.

The raw ADC count is converted to units of energy in MeV by scaling the total kinetic energy of the events to match the average of four literature values for ^{235}U [7–10]. A small linear volume-to-volume correction is applied based on the positions of the light and heavy fragment peaks. As a crosscheck, the total kinetic energy found in the NIFFTE TPC data versus neutron energy is compared against three literature values [10, 12, 13]. The mean value most closely matches those from GEF simulation [12], but the energy resolution is found to be greater than 13 MeV, which is several MeV worse than other measurements.

A full description of the "2E Method" for determining fission product yields can be found in a separate talk at this workshop [11]. Aside from the energy corrections, the mechanics of the iterative method are unchanged by the detector type and can be applied equally to the NIFFTE TPC. Preliminary results for ^{235}U fission product yield versus neutron energy are presented in Fig. 2. The yields are comparable to the known England and Rider standard [14], but the mass resolution of these results were found to be greater than ~8 a.m.u., which is substantially broader.

The next stage of development for the 2E method with NIFFTE is to focus on improving the largest areas of systematic uncertainty. An improved model-based energy loss correction in material will be implemented. Effects of "pulse height defect", *i.e.*, ionization energy losses of any type in the detector, will be studied, and could be corrected using an approximation based on empirical values [15] or from ion beam measurements in fissionTPC drift gas. The uncertainty due to estimation of neutron evaporation in the fission process (common to the 2E method in all detectors) can be assessed by considering alternate $\nu(A)$ models. Finally, the tracking ability of the TPC can be specifically exploited by using accurately-measured track length in place of energy (which has worse resolution), and by considering ionization profile information.

4 Use of dE/dx profiles for improved fission fragment identification

As mentioned in Section 2, the TPC potentially gives us further track-related information, such as energy, length, and Bragg peak height and position. However, it is important to note that the general shape of the Bragg curve for fission products is quite different than that one might expect from α particles. We use the SRIM and TRIM software packages [16] to simulate and parameterize the energy loss curves expected for fission products passing through our gas volume. From other studies [17], we expect this to provide a reasonable agreement between the simulation and reality.

We consider two starting points based on SRIM when practically modeling the expected Bragg curves in our data: interpolation of SRIM look-up tables spanning a large range of A, Z, and E, and a parametrization using an empirical function fitted to SRIM outputs. In both cases, we estimate the amount of nuclear energy losses appearing as ionization based on a fit to TRIM simulations. Then we apply a data-determined offset and Gaussian smearing to these idealized curves to model how these curves would appear in our data. In order to

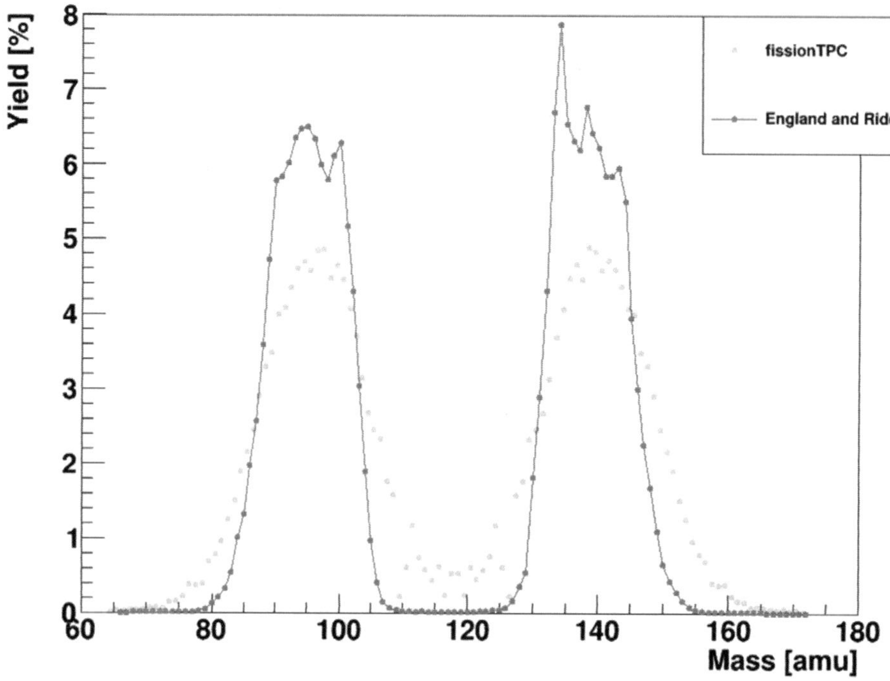

Figure 2. Preliminary ^{235}U fission product yield measurements using the 2E method with NIFFTE TPC data for incident neutron energies of $0.18 < E_n < 0.20$ MeV. The green points are values derived from data, the red connected line represents the literature reference [14].

determine the value of A and Z for a given event, we minimize the χ^2 difference between the Bragg curve in the data and that from our model.

A representative Bragg curve fit result is shown in Fig. 3. Preliminary results on ^{235}U data using solely dE/dx Bragg curve information give qualitative indications for light and heavy fission fragments clustered in A and Z, but also clear indications of pathological and non-physical results (Fig. 4) that remain to be understood.

We intend several avenues of investigation in order to improve these results. Since the 2E Method can effectively provide a value for A (to which Bragg curve measurements are relatively insensitive), it can be used as constraint on the Bragg curve fit to provide a value for Z. The basis of the dE/dx curves could also benefit from improvements, *e.g.*, better understanding of tracking/ionization pulse height defect, alternative models to SRIM, or potentially restriction of the analysis to well-known regions of the curve. A more complicated but potentially useful approach could be to implement machine learning techniques for event and curve classification.

Machine learning techniques have been typically applied with great success in image identification applications. In this case we would use points along the Bragg curve distribution as inputs, with neural network training to output values of A and Z. If we cannot determine values of A and Z outright with machine learning, we might at least discover the most relevant regions or metrics for Bragg curve identification. Another possible application could be identification and removal of events with poor tracking, low detector efficiency, *etc.*

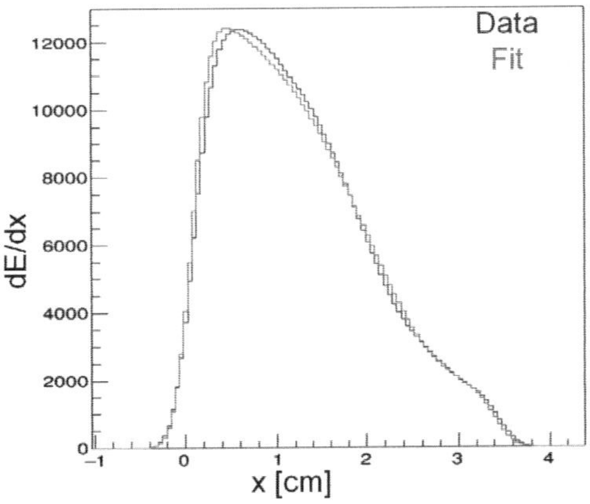

Figure 3. An illustrative example of a measured Bragg curve and the curve resulting from the fit minimization.

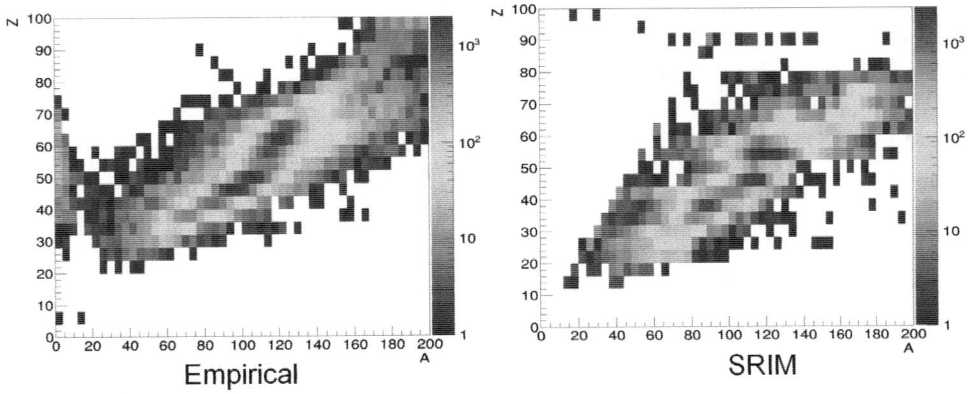

Figure 4. Preliminary ^{235}U yields of Z versus A from Bragg curve fits. The left plot uses an empirical estimate based on SRIM, while the right plot uses interpolated SRIM look-up tables. In both cases qualitative indications for heavy and light fission fragment regions are present, but are overshadowed by pathological and unphysical results.

In any case, one of the primary concerns with using machine learning is the ability to provide explainations for the decisions made, and to attribute uncertainties in a proper manner.

As a preliminary investigation, we have produced a neural network based upon Keras [18] and TensorFlow [19], which takes a toy model for Bragg curves using simple mathematical functions and outputs a decision based on training of four different sample curves. The model takes 20 points along the sample Bragg curve as input, with varied offset and scaling for each curve. Within this test exercise, we find a 99.7% classification accuracy. The next step is to generate a more realistic training data set using TRIM output convolved with detector resolution effects. This training can also include other variables, such as track length, and Bragg peak energy and location, in the list of inputs. An underlying principle is to understand the relative importance of each input node towards the final classification decision, in order to gain a greater understanding of the underlying physics information. This might be achieved through the use of software tools such as LIME (Local Interpretable Model-Agnostic Explanations) [20].

5 Conclusions

This talk has presented the starting phases of an analysis of ^{235}U fission product yields based on data collected by the NIFFTE collaboration fissionTPC. The NIFFTE detector provides complete 3D tracking information, including dE/dx, which is potentially useful for identification of A and Z mass numbers. Based on the iterative 2E method, working initial results consistent with previous measurements have been obtained, and focus has now shifted to understanding and reducing the associated uncertainties. We have begun to integrate ionization profile information into the data analysis, and are investigating the advantages of using machine learning techniques to further improve discriminatory power available by exploiting the full fission event information.

References

[1] R. J. Casperson *et al.* (NIFFTE Collaboration), Phys. Rev. C **97**, 034618 (2018).
[2] V. Geppert-Kleinrath *et al.* (NIFFTE Collaboration), Phys. Rev. C **99**, 064619 (2019).
[3] N. I. Walsh *et al.* (NIFFTE Collaboration), arXiv:1904.10558 (2019).
[4] M. Heffner *et al.* (NIFFTE Collaboration), Nucl. Instr. and Meth. A **759**, 50 (2014).
[5] M. Heffner, V. Riot, and L. Fabris, IEEE Trans. Nucl. Sci. **NS-60**, 2196 (2014).
[6] L. Snyder *et al.* (NIFFTE Collaboration), Nucl. Instr. and Meth. A **881**, 1 (2018).
[7] J. C. D. Milton and J. S. Fraser, Can. J. Phys. **40**, 1626 (1962).
[8] H. W. Schmitt, J. N. Neiler, and F.J. Walter, Phys. Rev. **141**, 1146 (1966).
[9] M. J. Bennett and W. E. Stein, Phys. Rev. **156**, 1277 (1967).
[10] D. L. Duke, Ph.D. thesis, LA-UR-15-28829 (2015).
[11] For a more complete discussion of the 2E Method, *e.g.* please see the talk by D. L. Duke at this workshop.
[12] K.-H. Schmidt *et al.*, Nucl. Data Sheets **131** 107 (2016).
[13] R. Yanez *et al.*, Nucl. Phys. A **970**, 65 (2018).
[14] T. R. England and B. F. Rider, LA-UR-94-3106 (1994).
[15] F.-J. Hambsch, J. van Aarle, and R. Vogt, Nucl. Instr. and Meth. A **361**, 257 (1995).
[16] J. F. Ziegler, M. D. Ziegler, and J. P. Biersack, Nucl. Instr. and Meth. B **268**, 1818 (2010).
[17] J. G. Hajnrych, P. Filliatre, and B. Geslot, Nucl. Instr. and Meth. A **911**, 15 (2018).
[18] F. Chollet *et al.*, https://keras.io (2015).
[19] M. Abadi *et al.*, http://tensorflow.org/ (2015).
[20] M. T. Ribeiro, S. Singh, and C. Guestrin, arXiv:1602.04938 (2016).

Late Prompt Fission Gamma Rays from ^{235}U(n,f) and ^{252}Cf(sf)

G. Rusev[1,*], *E. M.* Bond[1], *T. A.* Bredeweg[1], *A.* Couture[1], *J. M.* O'Donnell[1], *M.* Jandel[1,2], *A. E.* Lovell[1], *S.* Mosby[1], *C. J.* Prokop[1], *I.* Stetcu[1], *P.* Talou[1], and *J. L.* Ullmann[1]

[1]Los Alamos National Laboratory, Los Alamos, NM 87545, USA
[2]University of Massachusetts Lowell, Lowell, MA 01854, USA

Abstract. Two measurements of fission γ rays were performed with the DANCE and NEUANCE arrays using the reactions ^{235}U(n, f) and ^{252}Cf(sf). Utilizing the fast time response of the detectors and a method for estimating the accidental background, we obtained the energy spectrum of the late prompt fission γ rays as a function of the time since fission. The experimental results are compared with predictions of the code CGMF folded with GEANT4 simulations of the detector response.

1 Introduction

Prompt fission γ rays (PFG) are often generally considered to be the γ rays emitted by the fission fragments before β-decay occurs. Indeed, a large fraction of the PFG are emitted shortly after scission. Many of the fission fragments exhibit isomeric states. These states cause a delayed emission of some of the PFG by tens or hundreds of nano seconds or longer. We will call these γ rays "late PFG" to distinguish them from the PFG emitted very quickly (within 1 ns) and the β-decay delayed γ rays.

Measurement of the late PFG can provide information for the population of specific isomers and, respectively, the spin of the fission fragment, the γ-ray strength functions and level densities that govern its population. The late PFG from a given isomer enable an experiment that does not identify the fission fragments to access a specific isotope and benchmark the performance of the Hauser-Feshbach routine for it. In addition, measurement of the PFG multiplicity can test the assumed spin distribution of the fission fragments in the fission codes [1].

There is a recent progress in development of event-by-event statistical fission codes like CGMF [2] and FREYA [3]. The codes calculate the energies and multiplicities of all particles released after scission (fission fragments, γ rays, and neutrons). CGMF also uses all known isomeric states and assigns emission time for each γ ray. Predictions with CGMF for the late PFG energy spectrum and γ-ray multiplicity of ^{252}Cf(sf) and thermal (n, f) reactions on ^{235}U and ^{239}Pu were published in Ref. [4].

In this report, we present new experiments on the ^{235}U(n, f) and ^{252}Cf(sf) reactions with the Detector for Advanced Neutron Capture Experiments (DANCE). We compare the energies of the measured late PFG as a function of the time since fission occurred with CGMF results folded with the DANCE detector response.

*e-mail: rusev@lanl.gov

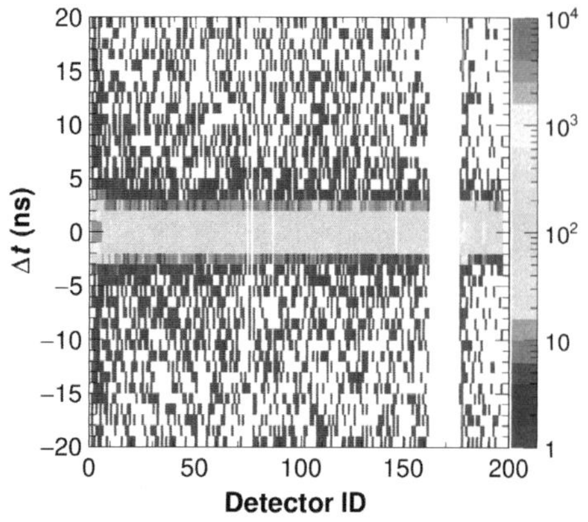

Figure 1. Time difference between the first detector of DANCE and any other of the DANCE and NEUANCE detectors. The DANCE detectors have IDs from 0 to 161 and the NEUANCE detectors from 176 to 196. A window of 5 ns can be used to build coincidences between all detectors.

2 Experiments, data analysis, and results

Measurements of the late PFG from $^{235}U(n, f)$ and $^{252}Cf(sf)$ were carried out with the DANCE array at the Manuel Lujan Jr. Neutron Scattering Center of the Los Alamos Neutron Science Center (LANSCE) [5]. We used a 26 mg/cm^2 thick ^{235}U target with a beam of white-spectrum spallation neutrons [6] in the first experiment and a 0.5 μCi ^{252}Cf source in the second one.

DANCE is constructed from 160 BaF_2 detectors in a 4π configuration to serve as a γ-ray calorimeter. The shape of the detectors is chosen such that they fit together without gaps [7]. In addition, the BaF_2 crystals are wrapped with very thin films to minimize the energy lost when a γ ray scatters between two detectors. In the standard configuration of DANCE, a ^6LiH shell is installed in the central cavity of DANCE, serving as an absorber for neutrons scattered from the target. In our experiments, the ^6LiH shell was replaced by NEUANCE [8] to provide a fission trigger. NEUANCE consists of 21 stilbene detectors arranged cylindrically around the target. A measurement of a prompt fission neutron (PFN) with NEUANCE is an indication that a fission event occurred.

Both detector arrays, DANCE and NEUANCE, provide fast response times enabling building coincidence events with a 5-ns window. Figure 1 shows the time difference between the first DANCE detector and any of the other DANCE and NEUANCE detectors.

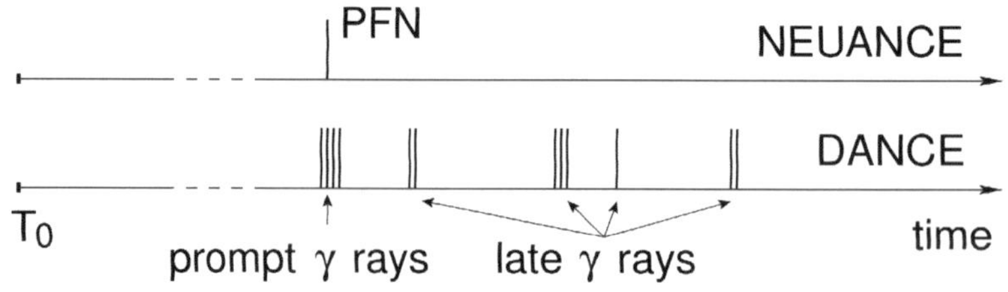

Figure 2. A schematic train of events in DANCE and NEUANCE with time starting from the beginning of the neutron spill T_0.

In the train of measured events, a fission event starts with a coincidence between a NEUANCE detector that registered a neutron and a few DANCE detectors fired by PFG. DANCE continues to detect γ rays with small multiplicity later in time. Most of these γ rays originate from background events, but some of them could be late PFG. Figure 2 shows a

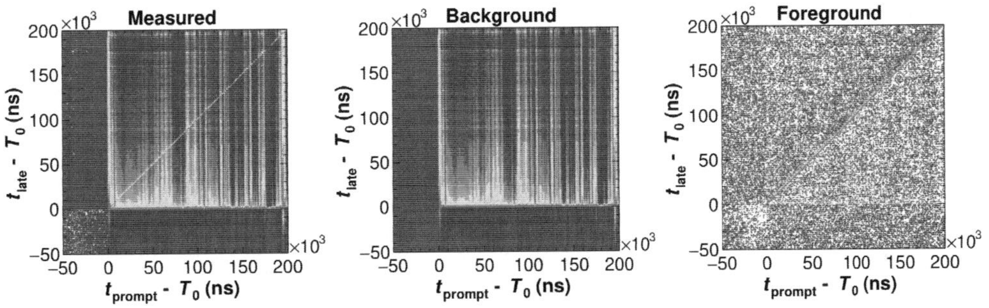

Figure 3. Measured t_{late} vs t_{prompt} spectrum (left), the estimated accidental background (middle), and the residuum of both (right).

schematic representation of the train of events in DANCE and NEUANCE. T_0 is the beginning of the spill of spallation neutrons.

Some of the background events are generated by the DANCE detectors. The α-decay of ^{226}Ra, a natural contaminant of barium, produces signals up to 3.5 MeV. An effect of the electronics, known as re-triggering, leads to double counting of a γ ray, where the second pulse height is artificial, giving much lower energy than the first pulse height. Both types of events can be easily removed by applying cuts in two-dimensional graphs, see *e.g.* Ref. [9] for more details.

Cosmic rays cause many of the DANCE detectors to fire due to their high γ-ray energies. This background can be filtered out by applying a multiplicity cut, because we expect the late PFG to have multiplicity close to one.

Other background events originate from the β-decay of naturally occurring isotopes like ^{40}K, ^{208}Tl, etc., or capture of PFN or beam neutrons scattered from the target in structural material of DANCE after some moderation time. These background events have similar characteristics as the late PFG, low multiplicity and low γ-ray energies, and cannot be removed by applying cuts on the total γ-ray energy of DANCE or γ-ray multiplicity, for example.

We applied a method for estimating the accidental coincidences to remove the natural and beam-related background. The method was originally developed for Chi-Nu experiments and is described in Ref. [10]. The method is based on the formula for the rate of the accidental background when measuring coincidence events with two detectors as the product of the rates of the two detectors and the coincidence window if dead time of the detectors is not considered. Written in terms of counts $r = ab/N \pm \sqrt{ab(a+b)/N^2}$, where a and b are the counts from the two detectors and N is the number of measurements. We apply the method by simplifying the multidetector arrays DANCE and NEUANCE to two type of events (i) prompt γ rays, which are the PFG, and (ii) late γ rays, which are the late PFG and background events (*cf.* Fig. 2).

We collect three time spectra: "t_{prompt}" - PFG relative to T_0, "t_{late}" - late γ rays relative to T_0, and "$t_{late} - t_{prompt}$" - late γ rays relative to the time when fission occurred. Multiplying the t_{prompt} and t_{late} spectra bin wise according to the formula $C_{prompt}(t_{prompt}) \times C_{late}(t_{late})/N_{T_0}$ creates a matrix of the accidental background (see Appendix B of [10] for more detail). $C_{prompt}(t_{prompt})$ is the counts in the PFG spectrum at a time t_{prompt}, $C_{late}(t_{late})$ is the counts in the late γ rays spectrum at a time t_{late}, and N_{T_0} is the number of neutron spills in the experiment. The background matrix is shown in Fig. 3 together with the measured two-dimensional histogram of t_{late} vs t_{prompt}. The diagonal elements for $t_{late} = t_{prompt}$ corresponds to the PFG, i.e. the time when fission occurred. The diagonal above corresponds to the late

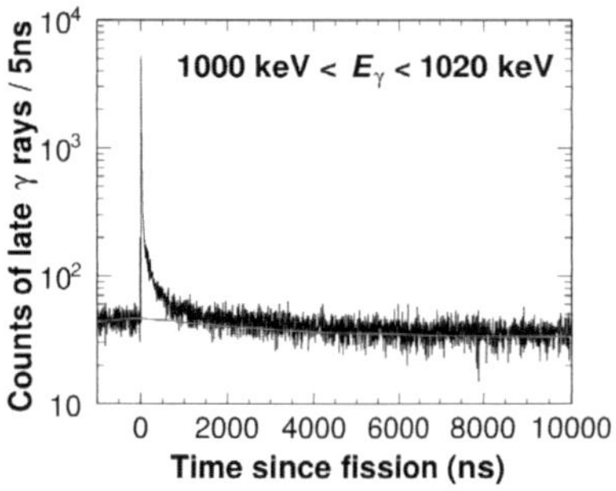

Figure 4. Measured $t_{late} - t_{prompt}$ spectrum for γ rays between 1000 and 1020 keV compared with the calculated background. Fission occurred at $t_{late} - t_{prompt} = 0$.

PFG that appear between 0 and 5 ns after fission. This is the diagonal with the strongest intensity visible in the measured spectrum in Fig. 3. Analogously, the next diagonal above corresponds to the late PFG between 5 and 10 ns since fission, etc. Summing the counts in each of the diagonals gives us the estimated accidental background for the $t_{late} - t_{prompt}$ spectrum.

In order to build a γ-ray energy spectrum, we require the measured γ rays to have energy within a given range with binning of 20 keV/channel, for example. We collect the t_{prompt}, t_{late}, and $t_{late} - t_{prompt}$ spectra for each bin of 20 keV and calculate the accidental background. An example $t_{late} - t_{prompt}$ spectrum is shown in Fig. 4. The counts in the $t_{late} - t_{prompt}$ spectra above the accidental background at a given time since fission produce a γ-ray energy spectrum of the late PFG.

Combining all γ-ray spectra gives the time evolution of the late PFG. The measured late PFG from the ^{235}U(n, f) reaction are shown in Fig. 5 and the results from ^{252}Cf(sf) are given in Fig. 6. The experimental results are compared with CGMF calculations for the two reactions folded with the DANCE detector response. A new GEANT4 model of DANCE and NEUANCE was created [11] to take into account the specifics of the detector setup utilized here. The model uses the measured γ-ray resolution and threshold of the individual detectors to ensure correct comparison between experiment and theory.

Examining closely the measured and calculated spectra in Figs. 5 and 6, we see a good similarity for the long-lived isomers. The energy resolution of DANCE smears γ rays with close energies that deexcite various isomers, emphasizing some of the strongest populated longer-lived isomers. We notice that the experimental two-dimensional spectra exhibit higher intensity at <50 ns than the CGMF spectra, suggesting existence of short-lived isomers with $T_{1/2} < 10$ ns not observed in previous experiments and, respectively, not included in CGMF.

To have a more quantitative comparison between CGMF and the experiment, we integrated the two-dimensional spectra from 50 ns to 2 μs. The integration started from 50 ns since fission to omit the γ rays from the short-lived isomers not included in CGMF. The γ-ray spectra are scaled to unity and shown in Figs. 5 and 6. There is an overall very good agreement between the CGMF results and the experiment. However, CGMF shows higher population of some isomers from ^{252}Cf(sf) leading to higher intensity γ rays around 0.55 and 0.85 MeV. CGMF also underpredicts the late PFG intensity near 0.7 MeV from the ^{235}U(n, f) reaction.

3 Summary

We carried out two experiments with the DANCE and NEUANCE arrays on the ^{235}U(n, f) and ^{252}Cf(sf) reactions. Using NEUANCE as a fission trigger and having a large mass of the ^{235}U enabled us to observe the late PFG with good statistics. Applying a novel method for estimating the accidental background helped us to remove the background in the measured

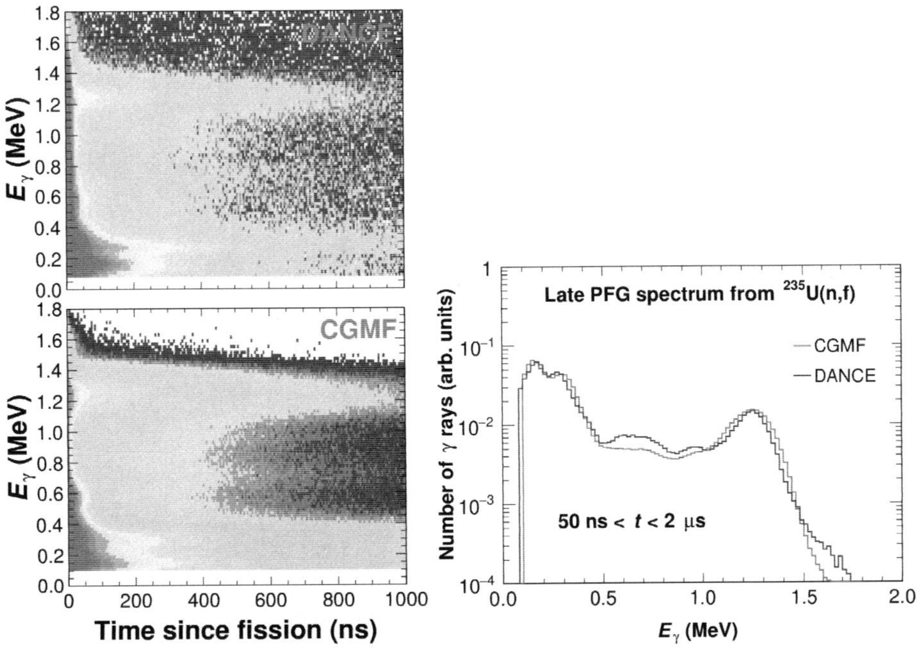

Figure 5. Measured γ-ray spectrum of the late PFG from the $^{235}U(n, f)$ reaction as a function of the time since fission (top-left panel). CGMF calculations for the same reaction folded with the DANCE detector response (bottom-left panel). Integrated experimental and theory spectra from 50 ns to 2 μs and normalized to unity (right).

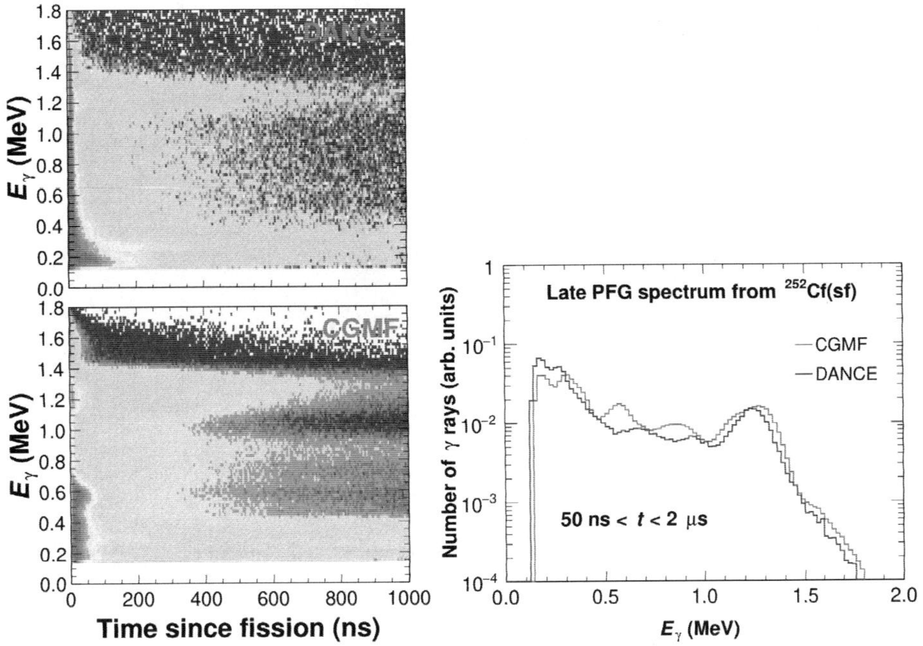

Figure 6. The same as Fig. 5, but for the late PFG from $^{252}Cf(sf)$.

spectra. We observed a good agreement of CGMF with the experiment with the exception of many short-lived isomers ($T_{1/2}$ < 10 ns) unknown from the literature. Comparing the integrated spectra of late PFG from 50 ns to 2 μs reveals that CGMF results closely resemble the late PFG from ^{235}U(n, f) and ^{252}Cf(sf) with small exceptions. In future work we will adress the discrepensies between experiment and theory.

4 Acknowledgements

This work has been supported by the Office of Defense Nuclear Nonproliferation Research and Development, National Nuclear Security Administration, U. S. Department of Energy.

This work benefitted from the use of the LANSCE facility at the Los Alamos National Laboratory.

References

[1] I. Stetcu, P. Talou, T. Kawano, and M. Jandel, Nuclear Data Sheets **118**, 230 (2014).
[2] P. Talou, R. Vogt, J. Randrup *et al.*, Eur. Phys. J. A **54**, 9 (2018).
[3] J. Randrup and R. Vogt, Phys. Rev. C **80**, 024601 (2009).
[4] P. Talou, T. Kawano, I. Stetcu *et al.*, Phys. Rev. C **94**, 064613 (2016).
[5] P. W. Lisowski and K. F. Schoenberg, Nucl. Instrum. and Methods A **562**, 910 (2006).
[6] M. Mocko, G. Muhrer, Ch.T. Kelsey *et al.*, Nucl. Instrum. and Methods A **632**, 101 (2011).
[7] M. Heil, R. Reifarth , M.M. Fowler *et al.*, Nucl. Instrum. and Methods A **459**, 229 (2001).
[8] M. Jandel, B. Baramsai, T.A. Bredeweg *et al.*, Nucl. Instrum. and Methods A **882**, 105 (2018).
[9] G. Rusev, LANL Technical Report **LA-UR-18-29064** (2018).
[10] J. M. O'Donnell, Nucl. Instrum. and Methods A **805**, 87 (2016).
[11] G. Rusev, LANL Technical Report (2019), in preparation.

Comparing fission-product yields from photon-induced fission of ^{240}Pu and neutron-induced fission of ^{239}Pu as a test of the Bohr hypothesis in nuclear fission

Jack Silano[1,*], *Anton* Tonchev[1,2], *Roger* Henderson[1], *Nicolas* Schunck[1], *Werner* Tornow[2,3], *Calvin* Howell[2,3], *FNU* Krishichayan[2,3], *Sean* Finch[1,2,3], and *Matthew* Gooden[4]

[1]Nuclear and Chemical Sciences Division, Lawrence Livermore National Laboratory, Livermore, CA 94550, USA
[2]Duke University, Department of Physics, Box 90308, Durham, NC 27708-0308, USA
[3]Triangle Universities Nuclear Laboratory, Durham, NC 27708, USA
[4]Los Alamos National Laboratory, Los Alamos, NM 87545, USA

Abstract. Fission product yields (FPYs) are a uniquely sensitive probe of the fission process, with well established dependence on the species of nucleus undergoing fission, its excitation energy and spin. Thus FPYs are well suited for testing Bohr's hypothesis in the context of nuclear fission, which states that the decay of a compound nucleus with a given excitation energy, spin and parity is independent of its formation. Using FPYs, we have performed a new high-precision test of the combined effects of the entrance channel, spin and parity on the fission process from two of the most commonly used particles to induce fission - neutrons and photons. The ^{239}Pu(n,f) reaction at E_n = 4.6 MeV and the ^{240}Pu(γ,f) reaction at E_γ = 11.2 MeV were used to produce a ^{240}Pu* compound nucleus with the same excitation energy. The FPYs from these two reactions were measured using quasimonoenergetic neutron beams from the TUNL's FN tandem Van de Graaff accelerator and quasimonenergetic photon beams from the High Intensity γ-ray Source (HIγS) facility. The FPYs from these two reactions are compared quantitatively for the first time.

1 Introduction

Since the discovery of nuclear fission in 1938, there has been extensive efforts to develop a theoretical description of the fission mechanism. The compound reaction-channel formalism of Bohr and Wheeler, first proposed in 1939 [1] is remarkably successful at describing the fission process. An essential assumption of this theory is that the nucleus proceeds through the fission process in a well defined quantum state. Thus the state of the nucleus influences the fission process, while the formation mechanism of the compound nucleus does not. Consequently all fission theory relies on the "Bohr hypothesis", which states that the decay of a compound nucleus for a given excitation energy, spin, and parity is independent of its formation [2]. The Bohr hypothesis is a particularly reasonable assumption for nuclear fission, where the timescale for the formation of the compound nucleus ($\sim 10^{-16}$ seconds) is long compared to the time needed to reach an equilibrium. However, despite the Bohr hypothesis

*e-mail: silano1@llnl.gov

underlying nearly every modern theoretical fission model, it has yet to be tested experimentally to a high degree of precision. Currently, the Bohr hypothesis is verified to about 20% accuracy in fission using nuclear reactions such as (t,pf), (^{3}He,df), and (^{3}He,tf) [3–7]. In this work we propose to test Bohr's hypothesis in the context of nuclear fission by comparing the fission product yields (FPYs) from a compound ^{240}Pu* nucleus produced via two different nuclear reactions: ^{239}Pu(n,f) and ^{240}Pu(γ,f).

2 Experiment

In this work we performed the first-ever experiment to identify the practical consequences of the Bohr hypothesis in the context of induced nuclear fission. We measured the FPYs from the ^{240}Pu(γ,f) reaction at the HIγS facility at Triangle Universities Nuclear Laboratory (TUNL) [8] and compared the results to the FPYs obtained from the ^{239}Pu(n,f) reaction that our group recently measured at the TUNL Tandem Van de Graaff accelerator [9–12]. Both reactions produced a ^{240}Pu compound nucleus with the same excitation energy since the energies of the incident neutron and photon beams were purposely chosen to account for the 6.5 MeV difference in binding energy between ^{239}Pu and ^{240}Pu. The only difference was the spin-parity distributions from n + ^{239}Pu and $J^{\pi} = 1^{-}$ from γ + ^{240}Pu, which is widely believed to be inconsequential at high excitation energies.

2.1 Dual-fission chambers

One of the most significant components in measurements of fission product yields is determining the number of fission reactions that occurred in the target. In the present work the fission reactions are counted using a Dual-fission ionization Chamber (DFC) [9], based on the design by Grundl *et al.* [13] for FPY measurements in reactors and critical assemblies [14].

The irradiation for the ^{239}Pu(n,f) measurements in Gooden *et al.* [11] were performed with the thick ^{239}Pu activation target mounted in the center of the DFC, as is the standard procedure used by our group [9–11, 15]. The two FCs then measure the effective neutron flux just before and after the target, and the effects of the beam divergence can be accounted for with Monte Carlo calculations [9]. The previously published data for ^{239}Pu(n,f) at E_n = 4.6 MeV were only sensitive to FPYs with half lives of ~days or longer, so the data were augmented with a new recent measurement using the same experimental procedure and target but with only 2 hours of irradiation, in order to measure the yields of FPYs with half lives of a few minutes to a few hours.

In the case of the ^{240}Pu photofission measurements, the thick ^{240}Pu activation targets were positioned just outside of the DFC, with the long-activation target upstream and the short-activation target downstream with respect to the γ-ray beam. Unlike the neutron beams at TUNL, the HIγS γ-ray beam is highly parallel with negligible divergence over a few cm distance. Thus it is not necessary to account for the spread of the γ-ray beam between the FC and the targets. This assumption was confirmed by comparing the relative beam flux measured in each chamber in the DFC. When a DFC is mounted in the neutron beam at TUNL, the beam divergence causes the upstream FC to typically see ~ 25% greater flux than the downstream FC, after accounting for the masses of the reference foils in each chamber. In the HIγS γ-ray beam, the ^{240}Pu DFC upstream to downstream FC ratio of 1.02 ± 0.01 was reasonably consistent with negligible γ-ray beam divergence. In addition, the self-absorption and scattering of the 11.2 MeV γ-ray beam from the thick foils was less than 1%, as calculated by MCNP simulations.

2.2 Activation targets & reference foils

The ^{240}Pu targets used in this measurement both consist of $^{240}PuO_2$ powder which is compressed into a 1.27 cm diameter Al cylinder, the same diameter as the ^{239}Pu target that was used in Ref. [11]. The ^{240}Pu mass of each target was 73.01±0.04 mg and 89.11±0.05 mg, with both targets enriched to 99.873% ^{240}Pu. The Al holders have a mass of ~ 19 mg, with a 1.91 cm outer diameter and a 0.191 cm thickness. The Al material is relatively transparent to the γ-rays emitted by the β-decay of fission products and also is not activated by exposure to the 11.2 MeV γ-ray beam.

The reference foils for the DFC consist of ^{240}Pu electroplated onto a 1.27 cm diameter area on a Ti backing with ^{240}Pu masses of 25.97±0.18 μg and 23.60±0.12 μg and an isotopic enrichment of 99.873%. The spatial uniformity of the deposition was verified by scanning the surface of the foils with a 1 mm aperture α-detector. Like the ^{239}Pu reference foils, the ^{240}Pu deposits were sufficiently thin that fission fragments could easily escape the electroplated material and deposit their energy in the gas of the ionization chambers with nearly 100% efficiency.

2.3 Monoenergetic photon beams from the HIγS facility

The HIγS facility produces intense, quasimonoenergetic γ-ray beams via intra-cavity Compton backscattering of free electron laser photons and relativistic electrons [8]. Photofission measurements require intense γ-ray beams such as the ones provided by the HIγS facility since the (γ,f) reaction is typically a factor of ~5–10 smaller than the (n,f) reaction for a given target nucleus.

For this work, the HIγS facility was operated using 540 nm free electron laser photons to produce linearly-polarized, 11.2 MeV γ-ray beams with a resolution of approximately 360 keV (FWHM) and an average flux on target of 1.95×10^7 γ-rays/(cm^2 s). The γ-ray beam was collimated with a 1.27 cm diameter, 15.24 cm long Pb collimator placed ~2.5 m upstream of the target assembly. The energy spectrum of the γ-ray beam was measured with a 123% high-purity germanium (HPGe) detector positioned in the beam axis, while copper attenuators were inserted into the γ-ray beam far upstream of the collimator to reduce the flux.

One ^{240}Pu target (#1) was irradiated with E_γ=11.2 MeV γ-ray beam for 2 hours to measure the short-lived FPYs (minutes to hours), the other (#2) for 92.6 hours to build up activity of long-lived FPYs (days). The 2 hour irradiation induced a total of $8.25{\pm}0.15{\times}10^6$ fissions in the short-lived FPY target, while the 92.6 hour irradiation induced a total of $4.38 \pm 0.07 \times 10^8$ fission events.

2.4 Detection of γ-rays from the β-decay of fission products

After irradiation by the HIγS γ-ray beam, the activated ^{240}Pu targets were monitored by a 60% relative efficiency HPGe. The first target with 2 hours of irradiation was counted for ~ 90 hours while the second, long-lived FPY target was being irradiated. After the long irradiation was completed, the second ^{240}Pu target replaced the first one in the same HPGe counting station. Thus both ^{240}Pu targets were monitored by the same HPGe at the same distance. Since this HPGe detector is the same one that was used in the ^{239}Pu(n,f) measurements of Gooden *et al.* [11] (and in the supplemental ^{239}Pu(n,f) measurements made recently), all of the FPY measurements are on the same systematic footing. Consequently this work posseses a unique sensitivity for comparing FPY distributions from the ^{239}Pu(n,f) and ^{240}Pu(γ,f) reactions. The measured fission product γ-ray spectra from the ^{239}Pu(n,f) and ^{240}Pu(γ,f) reactions are compared in Fig. 1.

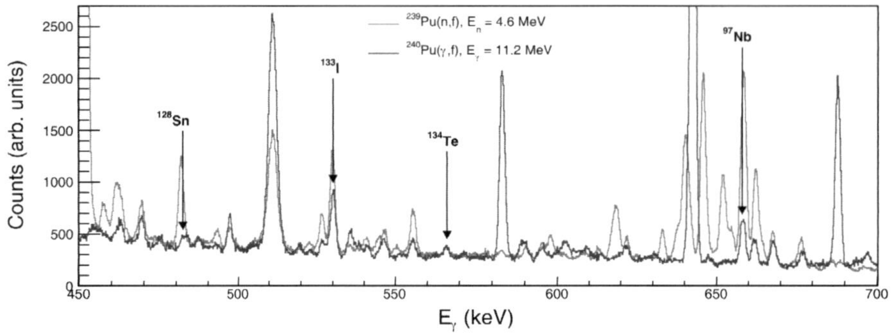

Figure 1. γ-ray spectra from fission products from ^{240}Pu(γ,f) at E_γ = 11.2 MeV and ^{239}Pu(n,f) at E_n = 4.6 MeV, measured by a 60% relative efficiency HPGe. Both spectra come from targets which were irradiated for 2 hours.

2.5 Analysis & data reduction

Yields of specific fission products are determined by counting γ-rays emitted by the decay of the corresponding fission products. Choosing the counting window in time is complicated by the interplay of the activity of the γ-ray of interest, interfering background peaks and the smooth continuum background. Because all of these features are time dependent, it can be difficult to predict a priori what the ideal integral time window is to fit the net photopeak area. In this work, a procedure was developed to automatically determine the integration time in a way which minimizes the uncertainty in the net photopeak counts, which is often one of the largest sources of uncertainty in the short-lived FPY measurements. Multiple spectra were generated with each successive spectrum having a longer counting time. The photopeak of interest was fit in each spectrum with a Gaussian function, including a linear background and other Gaussian peaks. The uncertainty of the net photopeak counts from the fit was then plotted as a function of counting time. In addition to minimizing the uncertainty in the photopeak counts, this method removes the choice of the counting period as a potential source of bias.

Using the same formalism as in Ref. [11], the FPYs can be defined as:

$$\mathrm{FPY}_i = \frac{\lambda_i N_i}{F_T n_a I_{\gamma i} \epsilon_i} \frac{1}{1 - e^{-\lambda_i t_e}} \frac{1}{e^{-\lambda_i t_d}} \cdot \frac{1}{1 - e^{-\lambda_i t_m}} \prod_k C_{ki}, \tag{1}$$

 https://doi.org/10.1051/epjconf/202024201008

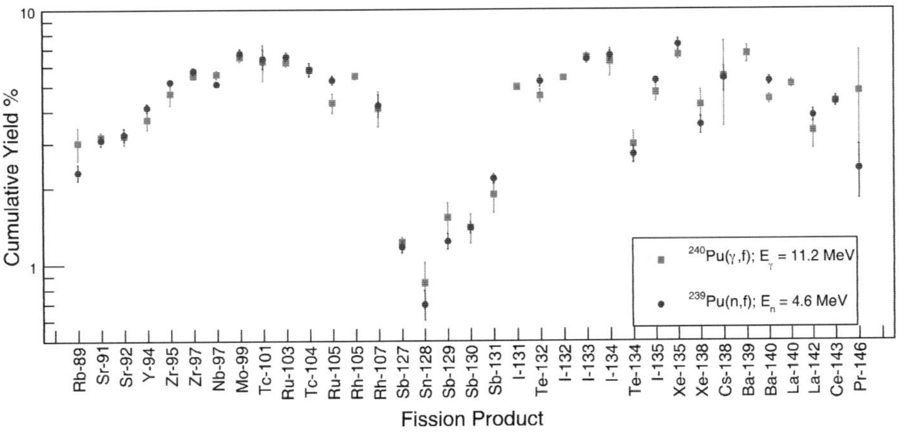

Figure 2. Preliminary FPYs from ^{240}Pu(γ,f) at E_γ = 11.2 MeV and ^{239}Pu(n,f) at E_n = 4.6 MeV. The ^{239}Pu(n,f) FPYs include new measurements of short-lived FPYs in combination with 13 long-lived FPYs from Gooden *et al.* [11]

where

$$
\begin{aligned}
\lambda_i &= \text{Decay constant} \\
N_i &= \text{Number of counts in the photopeak area} \\
n_a &= \text{Number of target nuclei} \\
I_{\gamma i} &= \text{Branching ratio} \\
\epsilon_i &= \text{Photopeak efficiency of HPGe detector} \\
F_T &= \text{Fission rate in target determined from the DFC} \\
t_e &= \text{Time of beam exposure} \\
t_d &= \text{Decay time from end of activation} \\
t_m &= \text{Target measurement time} \\
C_{ki} &= \text{Correction factors}
\end{aligned}
\tag{2}
$$

The correction factors include effects due to beam fluctuations, γ-ray attenuation and summing in the HPGe.

3 Results and discussion

The 35 unique FPYs measured from ^{240}Pu(γ,f) at E_γ = 11.2 MeV are plotted in Fig. 2 along with the corresponding yields from ^{239}Pu(n,f) at E_n = 4.6 MeV, if available. There were 29 unique FPYs which were measured for both sets of data. The FPYs are generally in good agreement, though there are some notable differences where the yields differ by multiple standard deviations. Significantly, the light mass peak in the FPY distribution doesn't show a systematic shift between the two distributions. Such a shift is expected in the case where the fissioning compound nucleus has a different mass, as the heavy fragment tends to remain constant due to shell closures, and additional nucleons are added to the light fragment. The lack of a systematic difference between the two FPY distributions is consistent with Bohr's hypothesis.

In order to quantify the similarity between the two sets of FPYs, a goodness of fit value may be defined as

$$\chi^2 = \sum_i \frac{\left(Y_{^{240}Pu(\gamma,f)}(^AZ_i) - Y_{^{239}Pu(n,f)}(^AZ_i)\right)^2}{\sigma_{^{240}Pu(\gamma,f)}{}^2 + \sigma_{^{239}Pu(n,f)}{}^2}, \qquad (3)$$

where $Y_{^{240}Pu(\gamma,f)}(^AZ_i)$ and $Y_{^{239}Pu(n,f)}(^AZ_i)$, are the yields of the fission product ^AZ_i from $^{240}Pu(\gamma, f)$ and $^{239}Pu(n, f)$, respectively; and $\sigma_{^{240}Pu(\gamma,f)}$ and $\sigma_{^{239}Pu(n,f)}$ are the uncertainty in the ^{240}Pu(γ,f) and ^{239}Pu(n,f) FPYs, respectively. The number of degrees of freedom is equal to the number of pairs of FPYs. A histogram of the χ^2 for each individual FPY pair is shown in Fig. 3.

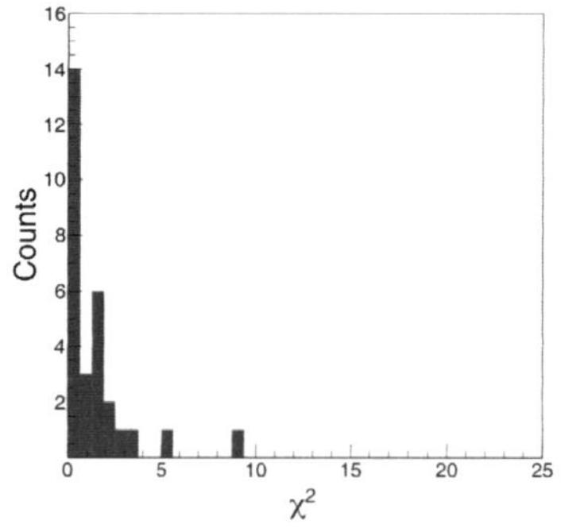

Figure 3. Preliminary χ^2 distribution for the comparison of FPYs from ^{240}Pu(γ,f) at E_γ = 11.2 MeV and ^{239}Pu(n,f) at E_n = 4.6 MeV.

The total χ^2 per degree of freedom for this work is 1.36, reasonably consistent with the expected value of 1 if the FPY distributions from ^{240}Pu(γ,f) and ^{239}Pu(n,f) at the same compound nuclear excitation energy are the same. Thus, these preliminary results are consistent with Bohr's hypothesis.

4 Conclusion & Future Work

This work represents the first quantitative comparison of FPYs from the ^{240}Pu(γ,f) and ^{239}Pu(n,f) reactions at the same compound nuclear excitation energy. The primary goal of the present experiments was to obtain an accurate comparison of the ^{240}Pu(γ,f) and ^{239}Pu(n,f) FPY measurements. The required accuracy was obtained by using the same experimental techniques and exact same HPGe detectors for both measurments, eliminating systematic errors which would otherwise make a direct comparison of the data sets difficult.

In order to improve the validation of Bohr's hypothesis in the context of fission, we have performed additional measurements of the FPYs from the ^{240}Pu(γ,f) reaction at E_γ = 8.0 and 11.2 MeV. The 11.2 MeV measurement was performed with significantly greater γ-ray beam flux, which will increase the number of measurable yields and decrease the statistical uncertainties on the FPYs presented in this work. The 8.0 MeV data set can be compared to existing ^{239}Pu(n,f) data at E_n = 1.5 MeV [11], allowing us to test if the energy evolution of the FPYs from ^{240}Pu(γ,f) and ^{239}Pu(n,f) are consistent. the E_γ = 8.0 MeV data, in conjunction with the E_n = 1.5 MeV data, should also be more sensitive to the pairing and shell effects which are more pronounced at these relatively low excitation energies. In this case the small mismatch of the angular momentum in the entrance channel will be further investigated.

5 Acknowledgements

We would like to thank the HIγS staff for delivering excellent quality γ-ray beams for this work. This work was performed under the auspices of the U.S. Department of Energy by Lawrence Livermore National Laboratory under Contract DE-AC52-07NA27344. This work was also supported partially by the U.S. Department of Energy, Office of Nuclear Physics, under Grant No. DE-FG02-97ER41033, and by the National Nuclear Security Administration under the Stewardship Science Academic Alliance Program through the U.S. Department of Energy, Grants No. DE-NA0002793, DE-NA0002936, DE-NA0003884 and DE-NA0003887.

References

[1] N. Bohr, Phys. Rev. **56**, 426 (1939)
[2] N. Bohr, Nature **137**, 344 (1936)
[3] H.C. Britt, F.A. Rickey, W.S. Hall, Phys. Rev. **175**, 1525 (1968)
[4] J.D. Cramer, H.C. Britt, Phys. Rev. C **2**, 2350 (1970)
[5] H.C. Britt, J.D. Cramer, Phys. Rev. C **2**, 1758 (1970)
[6] W. Younes, H.C. Britt, Phys. Rev. C **67**, 024610 (2003)
[7] W. Younes, Phys. Rev. C **68** (2003)
[8] H.R. Weller, M.W. Ahmed, H. Gao, W. Tornow, Y.K. Wu, M. Gai, R. Miskimen, Prog. Part. Nucl. Phys. **62**, 257 (2009)
[9] C. Bhatia, B. Fallin, M. Gooden, C. Howell, J. Kelley, W. Tornow, C. Arnold, E. Bond, T. Bredeweg, M. Fowler et al., Nucl. Instrum. Methods Phys. Res., Sect. A **757**, 7 (2014)
[10] C. Bhatia, B.F. Fallin, M.E. Gooden, C.R. Howell, J.H. Kelley, W. Tornow, C.W. Arnold, E. Bond, T.A. Bredeweg, M.M. Fowler et al., Phys. Rev. C **91**, 064604 (2015)
[11] M. Gooden, C. Arnold, J. Becker, C. Bhatia, M. Bhike, E. Bond, T. Bredeweg, B. Fallin, M. Fowler, C. Howell et al., Nucl. Data Sheets **131**, 319 (2016)
[12] A. Tonchev, M. Stoyer, J. Becker, R. Macri, C. Ryan, S. Sheets, M. Gooden, C. Arnold, E. Bond, T. Bredeweg et al., *Energy Evolution of the Fission-Product Yields from Neutron-Induced Fission of 235U, 238U, and 239Pu: An Unexpected Observation*, in *Fission and Properties of Neutron-Rich Nuclei* (WORLD SCIENTIFIC, 2017), ISBN 9789813229426
[13] J.A. Grundl, D.M. Gilliam, N.D. Dudey, R.J. Popek, Nucl. Technol. **25**, 237 (1975)
[14] H. Selby, M.M. Innes, D. Barr, A. Keksis, R. Meade, C. Burns, M. Chadwick, T. Wallstrom, Nucl. Data Sheets **111**, 2891 (2010)
[15] C. Bhatia, B. Fallin, C. Howell, W. Tornow, M. Gooden, J. Kelley, C. Arnold, E. Bond, T. Bredeweg, M. Fowler et al., Nucl. Data Sheets **119**, 324 (2014)

Prompt Fission Gamma-Ray Measurements at UML Research Reactor

Razvan Stanescu[1] , *Hadrick* Green[1], *Toby* Morris[1], *Gencho* Rusev[2], and *Marian* Jandel[3]

[1]Physics and Applied Physics PhD candidate, University of Massachusetts-Lowell 01854, USA
[2]Los Alamos National Laboratory, Nuclear and Radiochemistry Group, Los Alamos 87545, USA
[3]Physics and Applied Physics Faculty, University of Massachusetts-Lowell 01854, USA

Abstract. Neutron-induced fission of ^{235}U was studied at the thermal column of the UMass Lowell 1 MW Research Reactor. A collimated, 2.25-inch diameter beam of thermal neutrons with the flux of ~5x10^5 n/cm^2/sec induced fission reaction on a plate of low-enriched uranium with the areal density ~25 mg/cm^2 of ^{235}U. We have used the prompt fission-neutron tagging method to identify the fission reaction in the off-line analysis. The method employs the pulse-shape discrimination of neutrons and gamma-ray events in stilbene scintillator and enables identification of coincidence events of prompt fission gamma-rays and prompt fission neutrons in coincidence time intervals less than 20-30 ns. The prompt gamma-ray radiation was detected using two co-linear NaI(Tl) detectors. The measured spectra of prompt-fission gamma rays between 150 keV and 6 MeV are presented. The results from these initial measurements demonstrate the feasibility of the experimental method. Future measurements with extended arrays of detectors are planned.

1 Introduction

Prompt fission gamma-ray (PFG) radiation represents approximately 3% of the total energy released from ^{235}U fission. Understanding of the PFG spectra and multiplicity distribution is critical for the development of future nuclear reactors and impacts the design of their shielding and cooling systems. The main part of the total heating in a non-fuel region of thermal reactors is due to gamma rays (up to 90%) [Bla07]. The PFG spectrum extends above 10 MeV. High energy PFG can penetrate through thick shielding materials. As such, PFG may provide useful signatures for the special nuclear material (SNM) detection, if the SNM is actively interrogated by neutron or high energy gamma-ray probes [Goz09] inducing fission. Studies of the PFG properties are also motivated by basic science to improve models of nuclear fission and statistical evaporation from compound nucleus. The total kinetic energy of fission fragments together with the prompt fission neutron spectra provide important information on the excitation energy of the primary fission fragments (PFF). Neutrons, however, do not carry away significant amount of angular momentum. The PFG, on the other hand, evaporate most of the angular momentum induced in PFF and we expect that the properties of PFG depend on the initial angular momentum distribution of PFF. At high excitation energies, neutron emission dominates PFG emission. However, the competition between neutrons and gamma rays is expected near the yrast band, where neutron emission is forbidden because of the high spin.

Fig. 1. Thermal column at UML Research Reactor with the removable beam extender.

The previous experimental work on prompt fission gamma includes data taken at Los Alamos [Jan18, Jan14, Jan12] and Budapest [Obe13] on prompt fission gamma spectra (PFGS) from ^{235}U thermal and resonance neutron-induced fission. The existing experimental data are compared to Cascade Gamma-Ray Multiplicity for Fission (CGMF) code simulations [Ste19]. The goal of this work is to establish a technique to measure prompt fission gamma rays with energies up to 20 MeV at the UML Radiation Laboratory, where different sources of neutrons are available and can be measured for thermal, fast and discreet neutron incident energy.

2 Experiments

The UMass Lowell Research Reactor (UMLRR) is a pool-type reactor currently licensed to operate at 1 MW. The reactor has various experimental facilities that make use of either reactor produced neutrons and gamma rays from high-activity ^{60}Co sources which are co-located in the reactor pool. In-core irradiation facilities provide a fission spectrum of neutrons and gamma rays, with total neutron flux up to 2×10^{13} n/cm^2/sec, and neutron and gamma ray dose rates on the order of 10^7 rads/hour. Ex-core irradiation facilities include a Fast-Neutron Irradiator that produces predominantly fast neutrons for material testing; a Pneumatic System for instrumental neutron activation analysis; a Thermal Column that provides predominately thermal neutrons for neutron radiography; three Beam Ports that can be used for a variety of sample and detector irradiations; and a Nitrogen-16 Irradiator that is used to calibrate dosimeters to high-energy gamma rays.

Our measurements are carried out at the thermal port of the UMLRR that provides neutrons in the thermal energy range for experimental application. The thermal column is comprised of two separate assemblies. One assembly, a 4 x 4-foot square column, is embedded within the reactor pool biological shield. The other assembly, the thermal column extension, is located between the pool liner and the nuclear core, and is supported by a structural member, which is firmly attached to the pool structure. The center line of the thermal column is aligned with the center line of the reactor core. The 4 x 4-foot square embedded column is filled with graphite. A neutron collimator for thermal neutron radiography is located in the center of the column. A lead gamma shield, located on the front end of the thermal column extension, provides shielding of reactor core gammas. 2 MeV fission neutrons travel through 4 feet of graphite and slow to thermal energies. Resulting energies correspond to wave lengths of approximately 0.9 - 6.4 Å. A heavy, steel thermal column door is provided as a shield to protect operation personnel against gamma radiation. Four 6” diameter access ports are provided in the face of the door. Each port is fitted with four separate boral-faced plugs. Each plug is drilled and tapped to accommodate a plug removal tool, which is used to insert

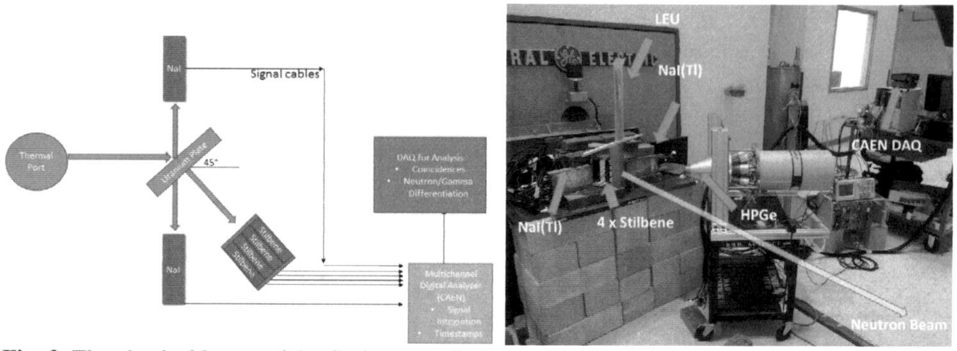

Fig. 2. The physical layout of the fission experiment.

and remove plugs. The maximum thermal neutron flux in the center of the column is ~10^{10} n/cm^2/sec with a cadmium ratio of ~100. At the outside face of thermal column (access port), the thermal neutron flux is ~1.5×10^5 n/cm^2/sec the cadmium ratio is ~7000. The neutron flux can be adjusted linearly down to 1.5×10^5 n/cm^2/sec.

The experimental setup is centered around the collimated 2.5" diameter thermal beam of neutrons from the reactor which is incident on the uranium plate angled at 45° as shown in Fig. 3 from above. The Low Enriched Uranium (LEU) plate is composed of ~20% ^{235}U and ~80% ^{238}U. The material is encased in an aluminum housing and measures approximately 24" x 3" x 0.15". The total thickness of ^{235}U in the fuel plate is ~25 mg/cm^2. Two NaI(Tl), four stilbene and HPGe detectors are used for the measurements. All detector signals are fed into a CAEN DT5730 digitizer which assigns time stamps and performs signal integration from stilbene and NaI(Tl) detectors with a 2-ns sampling interval. The DPP-PSD CAEN firmware enables FPGA real-time charge integration and asynchronous data acquisition for all channels.

Gamma rays were measured using two 2x2 inch NaI(Tl), placed ~2.625" from the U sample. The NaI(Tl) crystal has a density of 3.67 g/cm^3 and a decay time of 230 ns. This longer decay time puts a restraint on the maximum count rate in the detector without the pile-up. To counter this, both the distance from the detectors to the target and the reactor power level can be adjusted to control the flux of neutrons and reduce the reaction rates too acceptable level as well as the ambient background rate.

Fast neutrons were detected with stilbene scintillators. Four 1"x1"x4" stilbene detectors are arranged as a 4" x 4" wall placed 4.5" away from the target. Stilbene is an organic scintillator (trans - $C_{14}H_{12}$) with a density of 1.16 g/cm^3 which scintillates in different decay modes depending on the form of incoming radiation (in our case, neutrons vs gamma). Specifically, three decay components of the scintillation light in the stilbene crystal with the decay constant of <4 ns, ~18 ns, >35 ns [Zai09] can be categorized into "prompt" and "delayed" fluorescence. A ratio of the prompt and delayed fluorescence in a single event therefore enables incident radiation identification using the pulse shape discrimination (PSD). The ratio differs significantly for different incident radiation. Gamma rays induce significantly more prompt fluorescence compared to neutrons. With the fast response times and a decay time of ~4.5 ns [Leo94], stilbene detectors are ideal for coincidence measurements and are resolvable using a pulse shape discrimination 2-ns resolution CAEN digitizer. In addition, even the slowest component of the fluorescence light guarantees the signals to reach the baseline in a few hundred nanoseconds and enable high counting rate measurements. Because the stilbene-detectors gamma-ray energy resolution is not very good ~18% (consequence of the small photo-effect cross section in organic crystal), they are not used for spectral intensity measurements. Instead, we use them for identifying fission neutrons emitted in coincidence with gamma from higher resolution detectors as a tool for fission reaction tagging [Jan18]. The pulse shape discrimination consists of integrating the event's pulse shape, over time, in two different time intervals from the initial trigger of the pulse – the leading edge. Because of these different integration time lengths, they are referred to as a "short gate" and a "long gate". An appropriate way to visualize the two signal types is by plotting the following ratio compared to the long integral in Fig. 2:

$$\text{Ratio} = (\text{Long Integral - Short Integral}) / \text{Long Integral} \tag{1}$$

The coincident events were built in the off-line analysis to identify the detection of a fission event. In this manner, we were able to gate on the prompt fission neutrons in stilbene and identify coincident prompt fission gamma-rays on an event-by-event basis in the off-line data analysis.

Background measurements were taken in 3 different forms. Firstly, a background radiation measurement of the reactor at power but without the thermal neutron beam passing through. Secondly, a background measurement of the beam passing through the set-up but without the target in place, which allows us to determine the scattering of the radiation through the set-up materials. Finally, considering that the uranium target is encased in an aluminum casing, we used a similarly sized plate of aluminum, or a "dummy" target, to detect the scattering of the radiation off the target. Our results found the latter two background measurements to be nearly identical, and therefore satisfactory for background subtraction.

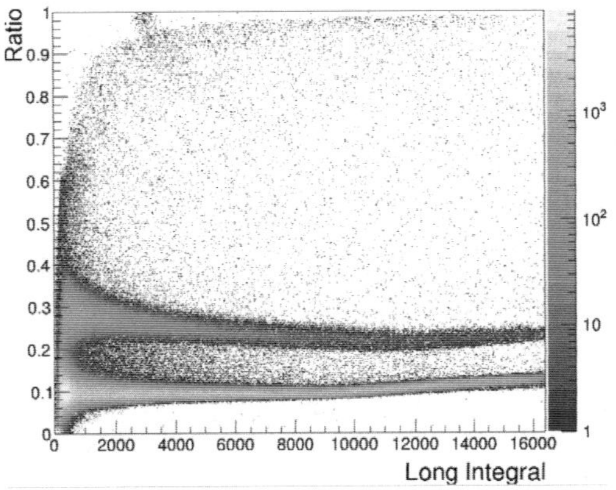

Fig. 3. Plotting the ratio from Eq. (2) versus the long integral of a signal, we can clearly see a delineation between the two radiation types. We can label each signal a neutron or a gamma based

3 Data Analysis

In the off-line data analysis, the detectors are first synchronized in time when time-offsets are added to each channel to guarantee the true coincidence. This is typically achieved with the ^{60}C source and confirmed with the data taken in beam. Energy calibration of all detectors is carried out as well. The data are then re-analyzed and coincident events are built in a 25-ns wide coincidence window. The coincident events that include neutron events from Stilbene detectors are then processed for detail PFG analysis. The width of the window is a variable and can be set to different values and as such, the dependence of PFG on time after the fission event can be studied. Due to the difference in response times of the NaI(Tl) and stilbene detectors as well as different length cables a very regular difference of timing was found between the two detector types. This consistent difference gives us confidence that the timing process in the digitizer and in our software, is well defined and reflects a replicable difference between the two detector type circuits. Because the signal timing shift was repeatedly found to be 17.5 ns as seen in Fig. 4, we could adjust the stilbene detectors by this time value so that coincident events are well correlated and we may narrow our coincidence window to be smaller around the collective Gaussian distribution and give us more finely sampled coincidence spectra.

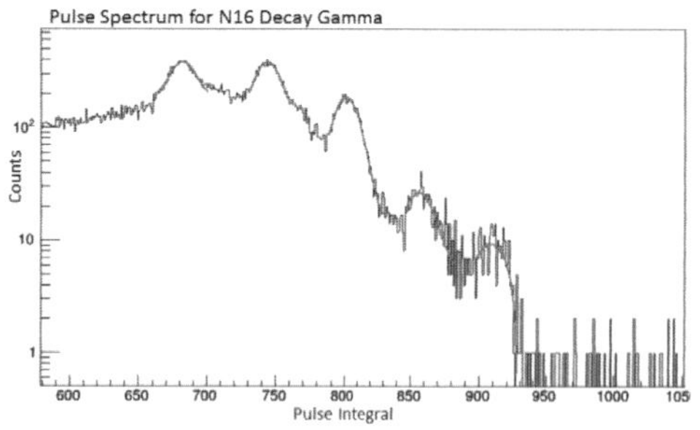

Fig. 4. Pulse spectrum for ^{16}N decay gamma. The five gamma peaks of a decaying ^{16}N ranging from 5.1 MeV to 7.1 MeV as detected in a NaI detector

Energy calibration of all detectors is obtained using the ^{137}Cs, ^{60}Co, and ^{133}Ba standard calibration sources. Over time however, it was found that the calibrations of NaI(Tl) would become inaccurate and would drift noticeably between the start and the end of a long day of data taking. We predict this is due either to the well documented variation in response that NaI(Tl) shows at different temperatures or due to systemic voltage variation in the power sources of the detector high voltage [Ian09]. To mitigate this effect, we found regular calibration before and after hours of data-taking to be sufficient in giving us accurate spectrum energies. As well, we explored a chi-squared method of comparing spectra of identical set ups with changing energy ranges over time. When comparing subsequent spectra to an original, well calibrated spectrum and scaling the energy domain as needed, the method allowed us to produce accurate calibrations over the course of many individual runs.

We found that at energies above 2 MeV, the linear calibrations would begin to deviate in a non-linear way. To appropriately calibrate this region, we used the pool of the reactor as a high energy gamma source. The neutrons coming from the core created ^{16}N by the following reaction with the oxygen in water: $^{16}O + n \rightarrow p + ^{16}N$. The ^{16}N nucleus beta decays with a half-life of $T_{1/2}$ =7.13 seconds with the dominant decay gamma rays with energies of 7.115 and 6.129 MeV and probabilities of 0.049 and 0.67 per decay, respectively [Kel93]. Due to the higher rate of pair production at energies over 1 MeV the peaks we expect will be accompanied by its two escape peaks below the main peak. Therefore, we expect to see 6 peaks in total, and due to one overlap we find the 5 peaks in Fig. 4 which we will use in future measurements to calibrate and characterize this non-linear energy region.

4 Results

Prompt fission gamma-ray spectrum from thermal neutron induced fission of ^{235}U is shown in Fig 5 by black open squares. The spectrum was obtained by gating on the neutron events detected in stilbene detectors in coincidence with fission gamma rays and coincidence window with of 25 ns. The background spectrum, obtained using the same offline data reduction procedure from the data taken with the Al dummy in neutron beam are shown in the same Fig 5 by open circles. Both spectra are normalized to 74 minutes of data acquisition. We can see that the background is a few orders of magnitude smaller compared to the prompt fission

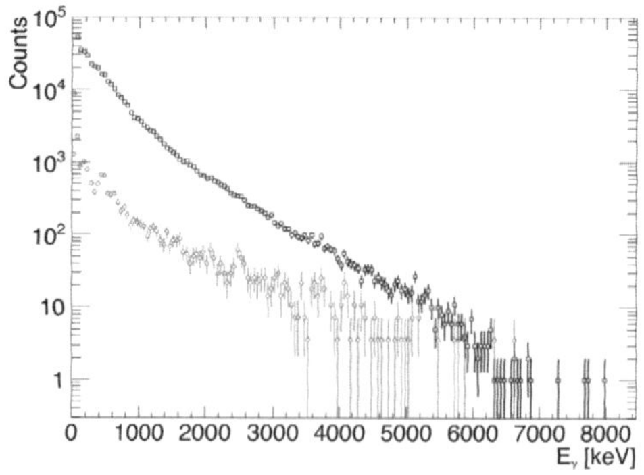

Fig. 5. Prompt fission gamma rays detected from ^{235}U thermal neutron-induced fission in ~2 hours-long measurement at the UML research reactor operating at 200 kW. The spectrum is obtained by gating on the prompt fission neutron events in stilbene detectors and for the coincidence window width of 25 ns. Background spectrum obtained from measurement with the Al plate is shown using open red squares.

gamma ray spectrum. This is encouraging for our future efforts to improve the precision and accuracy of the prompt fission gamma measurements.

5 Discussion and Future Work

We have successfully designed and carried out measurements of prompt fission gamma-rays in coincidence with the prompt fission neutrons in a collimated beam of thermal neutrons inducing the fission reaction on ^{235}U in a low enriched uranium fuel plate. The neutrons were detected using an array of stilbene scintillators and identified in the off-line pulse shape discrimination analysis. The gamma rays were detected in NaI(Tl) detectors. Coincident events were built in the offline data analysis with a selectable width of the coincident gate. The results presented in this work are shown for a 25-ns wide coincidence window. In future, we plan to significantly increase the counting statistics and run a week-long measurement in the summer of 2020. To validate our results, detectors of different type, such as BaF_2 will be used as well as larger volume NaI(Tl) detectors. We plan to utilize anti-Compton shields, available at UML, to surround 2-inch diameter detectors and actively reject the scattered gammas from the NaI(Tl) crystals. This will help improve the response of the detectors after removing the Compton background. Work is under way to field HPGe detectors in the same geometrical configuration and measure high-resolution spectroscopic data for fission gamma rays using the same fission neutron tagging method. Finally, the Geant4 and MCNP simulations will be carried out for our experimental set-up to better understand the scattering background and response function of the detectors. We would like to thank the staff of UML Research Reactor and Radiation Safety Office, for providing hours of reactor operations and support to field and run the experiment. We are indebted to Nuclear Radiochemistry Group at LANL for stilbene detectors of NEUANCE [Jan18] that were used in this measurement.

References

[Bla07] D. Blanchet et al., Annals of Nucl. Energy 35 (2007) 731-745.

[Goz09] T. Gozani, IEEE Transactions on Nucl. Sc., 56, 736 (2009).

[Jan18] M. Jandel et al., Correlated fission data measurements with DANCE and NEUANCE, *Nuclear Instruments and Methods in Physics Research Section A: Accelerators, Spectrometers, Detectors and Associated Equipment,* Volume 882 (2018) 105–113.

[Jan14] M. Jandel et al., Physics Procedia 59 (2014) 101–106.

[Jan12] M. Jandel et al., Los Alamos Report No. LA-UR-12-24975 (2012), to be published.

[Obe13] A. Oberstedt, T. Belgya, R. Billnert et al., Phys. Rev. C 87, 051602 (2013).

[Ste19] I. Stetcu. Evaluation of the Prompt Fission Gamma Properties for Neutron Induced Fission of $^{235,238}U$ and ^{239}Pu. (2019).

[Zai09] N. Zaitseva, et. al. Neutron detection with single crystal organic scintillators, SPIE Hard X-Ray, Gamma-Ray, and Neutron Detector Physics, 11 (2009).

[Leo94] W. Leo, *Techniques for Nuclear and Particle Physics Experiments, A How-To Approach*, Springer Science & Business Media, 159-164 (1994)

[Kel93] J. H. Kelley, D. R. Tilley, H. R. Weller and C. M. Cheves, NP 564 1 (1993). Data extracted using the NNDC On-line Data Service from the ENSDF database, revision of December, 1993.

[Ian09] K. D. Ianakiev, B. S. Alexandrov, P. B. Littlewood, and M. C. Browne, *Nuclear Instruments and Methods in Physics Research Section A: Accelerators, Spectrometers, Detectors and Associated Equipment,* Volume 607, Issue 211, Pages 432-438 (2009),

New Fission Product Yield Measurements at Oregon State University

Aaron Tamashiro[1,*], *Jason* Burke[2], *Stephen* Padgett[2], *Sean* Burcher[2], *Todd* Palmer[1], and *Camille* Palmer[1]

[1]Oregon State University
[2]Lawrence Livermore National Laboratory

Abstract. Cumulative fast fission product yields for ^{235}U and ^{238}U were measured at Godiva-IV in burst mode. Data was collected starting 45 minutes after the prompt irradiation. Data analysis codes were developed to analyze gamma-rays and calculate fission product yields with their respective uncertainties. Due to uncertainties in the branching ratios, different γ-rays from the same isotope led to different fission yield. This led to an effort at Oregon State University (OSU) to utilize the Rabbit facility at their 1 MegaWatt Training, Research, Isotopes, General Atomics (TRIGA) nuclear research reactor. The new experimental setup will allow us to measure branching ratio data and short-lived fission product yields starting seconds after a prompt irradiation.

1 Introduction

Fission product yields are used in the prediction of the behavior and performance of engineered systems involving nuclear fission. Measurements of fission product yields have been traditionally performed through radiochemistry, mass spectrometry, and γ-ray spectroscopy, often with mono-energetic neutrons. Recently, the Godiva-IV critical assembly was employed to perform a burst irradiation of a sample of ^{235}U and ^{238}U with the goal of measuring fission yields within an hour of prompt irradiation of neutrons in the Watt fission spectrum. An experiment at OSU will measure fission yields seconds after a prompt irradiation. We have developed a γ spectroscopy-based data analysis procedure that has been developed for the prediction of fission yields that is independent of the actinide sample for the Godiva-IV experiment and will be used for the OSU experiment.

2 Godiva-IV Experiment

Godiva-IV is a critical assembly composed of unreflected highly enriched uranium (HEU) metal fuel [1]. This device was intended to irradiate samples using prompt bursts of fast fission neutrons of around 24 μsec in duration with a peak neutron flux of around 4×10^{18} neutrons/cm^2/sec [1]. Large neutron fluxes allow shorter irradiation times, minimizing neutron absorption in the subsequent fission products, and assuring larger fission product generation rates. Counting on the sample began 45 min after irradiation.

*e-mail: tamashia@oregonstate.edu

Two Broad Energy Germanium (BEGe) detectors were used with a Lynx data acquisition (DAQ) system. The BEGe detectors were Canberra's model BE6530, which have a 60% relative efficiency. These were mounted on large cryostats filled with liquid nitrogen. The sample is placed 9 cm from each detector face. The γ-rays were detected by the BEGe spectrometers and were recorded in list mode format through the Lynx systems by Canberra's digital signal analyzer. The raw data is stored in separate files for each detector; the files contain a recorded time stamp and ADC channel for each γ-ray interaction. The detectors are synchronized, such that coincident events may be filtered out in the future, if needed. The resolution for the event clock timing was 100 ns and the resolution for the coincidence timing was 1 μs FWHM.

3 Data Analysis

The fission product yield, FPY_i, for isotope i is calculated directly from the ratio of the number of decay events from isotope i to the total number of fission events occured in an irradiation experiment. The total number of fission events N_f is calculated using assuming a known reference fission product yield, FPY_{ref}:

$$N_f = \left(\frac{A_0}{\epsilon_\gamma I}\right)\left[\frac{\tau_{1/2,ref}}{\ln(2) FPY_{ref} R_{ref}}\right]. \tag{1}$$

The measured fission yield for isotope i is written in terms of the reference isotope:

$$FPY_i = \frac{A_{0,i}\tau_{1/2,i}}{\ln(2) N_f \epsilon_{\gamma,i} I_i R_i} \tag{2}$$

Here, A_0 is the initial activity of the induced radioactive fission product, ϵ_γ is the absolute detector efficiency, I is the self attenuation of the γ-ray through the target, and R is the branching ratio of the γ-ray observed. γ-ray detection efficiency was measured using standard calibration sources with known activities. The self attenuation of γ-rays within the sample is measured using known γ-rays produced by the decay of the actinide. The γ-ray activities from the fission products are measured in the experiment, while the other parameters are taken from ENDF-VIII.0 [2]. The reference fission product used to calculate the total number of fission events is ^{99}Mo.

3.1 Interfering γ-Rays

To analyze the γ-ray peaks, references are needed to help identify which γ-ray peaks are from which decaying radioisotope. A reference library is compiled using data from ENDF-VIII.0 [2]. The compiled data assists the user in isolating the γ-ray of interest from other γ-rays nearby that would contribute into the activity of the γ-ray peak being analyzed.

3.2 γ-Ray Peak Analysis

The γ-ray peaks are analyzed with a code written in C++ and utilizing the ROOT data analysis framework [3]. By defining unique energy intervals for each peak, the peaks are isolated in the spectrum so they can be analyzed independently. These intervals are determined visually using cascade plots [see Figure 1].

Figure 1 shows an example of a cascade plots for a γ-ray peak at 196 keV associated with the fission product ^{88}Kr. Each color represents a different time bin and as time progresses the

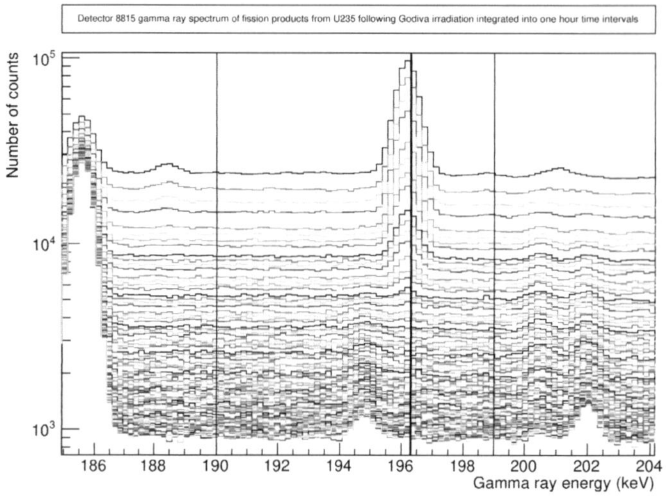

Figure 1. ^{88}Kr γ-ray cascade plot for the 196 keV γ-ray

overall count rate decrease. In this example, the data collected in the first three time bins is used to set the energy interval to [190.0 keV, 199.0 keV] to filter out the peaks near 188.5 keV and 201 keV. The interval chosen attempts to include as much background as possible; without including neighboring peaks.

As the time binned data is acquired, the live and raw time are used for the dead time correction by multiplying the counts for an energy bin with the ratio of raw time divided by live time. The ADC channel number is converted into γ-ray energy by using calibration sources. The time binned data are 2D histograms with E_γ on x-axis and time on the y-axis. Using this 2D histogram, a γ-ray spectrum for a specific time interval is produced by projecting the corresponding time bins on the x-axis. A Gaussian fit function with a linear background is used to fit peaks in the γ-ray spectrum:

$$F(x) = C \times \exp\left[-\frac{(x-\mu)^2}{2\sigma^2}\right] + mx + y, \tag{3}$$

where C is the amplitude of the peak, μ is the energy of the peak, σ is the resolution of the detector, m is the slope of the background, y is the y-intercept of the background, and x is energy. This technique is used when there is a background continuum present [4]. Before the code fits every time slice, a master fit is produced to provide the initial fit parameters for the γ-ray peak [see Figure 2]. The histogram for the master fit is the produced by summing the data over all the time bins in the chosen time interval for each channel.

The linear background is subtracted out and the integral S is the net counts in each respective time bin.

$$\begin{aligned} S &= \int \left(C \times \exp\left[-\frac{(x-\mu)^2}{2\sigma^2}\right] + mx + y\right) dx - \int (mx + y)\, dx \\ &= \int \left(C \times \exp\left[-\frac{(x-\mu)^2}{2\sigma^2}\right]\right) dx. \end{aligned} \tag{4}$$

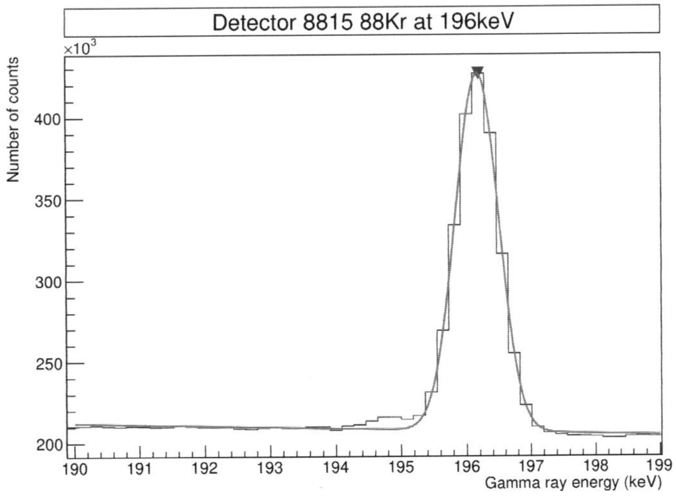

Figure 2. ^{88}Kr master sum plot fit for the 196 keV γ ray

3.3 Decay Curve Fit

Net counts and half-life data are used to fit a decay curve that is extrapolated to time zero. The data for this curve consists of the counts integrated over each peak (using the energy interval established previously) in each time bin . To secure a proper decay curve, the half life and time increments were fixed. The only free parameter in the fit was the initial count rate (A_0). Provided is the fit function:

$$N(t) = \left(\frac{A_0 \tau_{1/2}}{\ln(2)}\right)\left(exp\left(-\frac{\ln(2)}{\tau_{1/2}}t\right)\right)\left(1 - exp\left(-\frac{\ln(2)}{\tau_{1/2}}\Delta t\right)\right) \tag{5}$$

where $N(t)$ is the counts of γ-rays, A_0 is the initial count rate of γ-rays, $\tau_{1/2}$ is half life, Δt is time step, and t is time. This equation is derived from the integral of the count rate over a time bin.

Figure 3 is a plot of activity vs. time. The black points are the net counts with error for each time bin. The blue line is the fit function extrapolated over all time, the time interval used for the fit function is indicated by the highlighted thicker line on the plot.

In order to test the fit of the decay curve, a residual plot and χ^2_ν plot are created. Residual plots are created to visualize the relative difference between the decay curve fit and the net count data in the time interval chosen. The residual plot is another criteria used to help determine an energy and time interval for the γ-ray analysis. The time interval is varied until residuals are less than a user-defined value of 20%. The goal is to include as many points in the fit as possible, while minimizing the relative differences in the residual plot and keeping χ^2_ν close to 1.

3.4 Fission Product Yield Calculation

The initial count rate of γ-rays is passed on to a final code that uses Equation 2 to calculate fission product yield. The uncertainties were calculated using standard error propagation [5]. Different fission product yields from analyzing a γ-ray from a specific isotope compared to

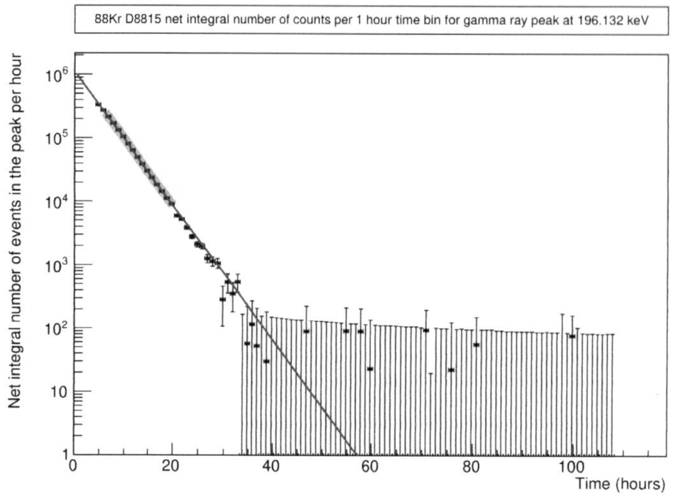

Figure 3. ^{88}Kr γ ray half-life plot for the 196 keV γ ray

another γ-ray from the same isotope would imply there are large uncertainties in the γ-ray branching ratio data or there are unknown interferences. Further information may be found in reference [6].

4 Oregon State University Test Run

A proof of principle measurement was done at OSU using their TRIGA Reactor Rabbit system. A sample of ^{238}U was transported into the reactor core pneumatically and irradiated for 2 seconds. The sample is then counted in list mode and the data was parsed into 30 second time bins [see Figure 4]. This allows for the measurement of short-lived ($\tau_{1/2} \leq 1hr$) fisison product yields. A future measurement is planned to evaluate independent fisison product yields and relative branching ratios.

5 Conclusions

The experimental setup was established in a room next door to the Godiva-IV critical assembly. The detection system and dimensions are all well established for future measurements. The data analysis routine is generalized and any modifications to the codes are easily applied. At least 100 γ-rays from around 30 isotopes were analyzed for ^{235}U and ^{238}U Even though the codes are generalized for any γ-ray peak in the spectra, human input is still needed for each γ-ray in the choice of energy window and time interval for the data analysis.

Future work will involve improvements in the data analysis. The improvements will eliminate human intervention and fully automate the data analysis. Also, the experimental setup and data analysis will be applied to other actinides to analyze their fission product yields. Improvements in the branching ratio data will be worked out with γ coincidence measurements at OSU. For Godiva-IV experiments, ^{239}Pu data is being analyzed and ^{237}Np is planned.

6 Acknowledgments

This work was funded by the Office of Defense Nuclear Nonproliferation Research and Development within the U.S. Department of Energy's National Nuclear Security Administration

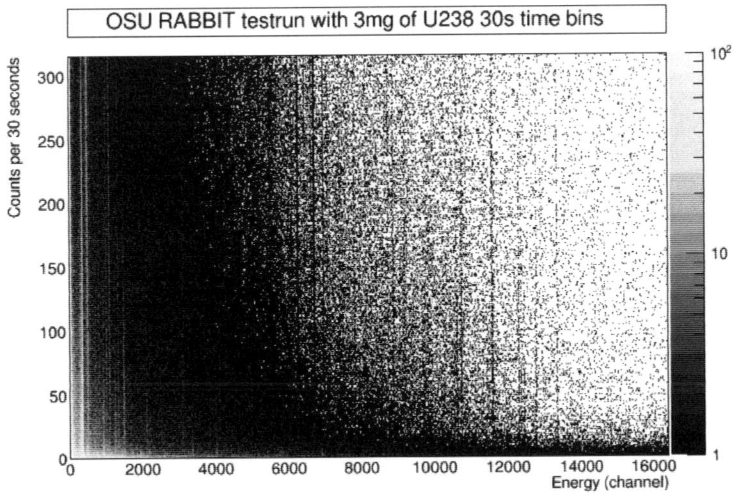

Figure 4. Test irradiation of U238 in the OSU TRIGA Rabbit

by Lawrence Livermore National Laboratory under Contract No DE-AC52-07NA27344. The U.S. Department of Energy's Nuclear Criticality Safety Programs National Criticality Experiments Research Center (NCERC), utilized in this work, is supported by the National Nuclear Security Administration's Office of the Chief of Defense Nuclear Safety, NA-511.

References

[1] R.E. Malenfant, Tech. rep. (1981)
[2] D.A. Brown et al., Nuclear Data Sheets **148**, 2887 (2018)
[3] I. Antcheva et al., Computer Physics Communication **180**, 2499 (2009)
[4] G.F. Knoll, *Radiation Detection and Measurement* (Wiley, 2010)
[5] P. Bevington, D.K. Robinson, *Data Reduction and Error Analysis for the Physical Sciences* (McGraw-Hill, 2010)
[6] A.S. Tamashiro, Master's thesis, Oregon State University (2019)

EPJ Web of Conferences **242**, 02001 (2020) https://doi.org/10.1051/epjconf/202024202001
FPY 2019

Present Status of Neutron-, Photo-induced and Spontaneous Fission Yields Experimental Data

B. Pritychenko[1,*], *O.* Schwerer[2], *J.* Totans[1], and *O.* Gritzay[2]

[1]National Nuclear Data Center, Brookhaven National Laboratory, Upton, NY 11973-5000, USA
[2]Under contract with National Nuclear Data Center, Brookhaven National Laboratory, Upton, NY 11973-5000, USA

Abstract. Nuclear reaction data collection, evaluation and dissemination have been pioneered at the Brookhaven National Laboratory since the early 50s. These activities gained popularity worldwide, and around 1970 the experimental nuclear reaction data interchange or exchange format (EXFOR) was established. The original EXFOR compilation scope consisted only of neutron reactions and spontaneous fission data, while many other nuclear data sets were ignored. Due to the high cost of new experiments, it is very important to find and recover the previously disregarded data using scientific publications, data evaluations and nuclear databases comparisons.
Fission yields play a very important role in applied and fundamental physics, and such data are essential in many applications. The comparative analysis of Nuclear Science References (NSR) and Experimental Nuclear Reaction (EXFOR) databases shows a large number of unaccounted experiments and provides a guide for the recovery of fission cross sections, yields and covariance data sets. The dedicated fission yields data compilation effort is currently underway in the Nuclear Reaction Data Centers (NRDC) network, and includes identification, compilation, storage and Web dissemination of the recovered data sets.

1 Introduction

The early success of Brookhaven National Laboratory (BNL) reactor program which generated large volumes of new results showed a strong need for data preservation and dissemination. Compilation of neutron cross sections have been initiated at BNL in the early 1950s by Donald Hughes (1915-1960). The Brookhaven group produced the famous BNL-170 report [1]. The Brookhaven report was a successor of the Metallurgical Laboratory, University of Chicago neutron cross section compilation [2, 3], and it is currently on display at the Donald Hughes book library as shown in Fig. 1. The National Nuclear Data Center (NNDC) book library is the major resource behind the NSR and EXFOR database contents development. Several nuclear reaction data projects such as the Atlas of Neutron Resonances reference book [4], Evaluated Nuclear Data File (ENDF) [5] and the Experimental Nuclear Reaction Data (EXFOR) ([6, 7] and references therein) predecessor and the SCISRS (Sigma Center Information Storage and Retrieval System) system [8] take their roots from the Brookhaven cross section compilations.

*e-mail: pritychenko@bnl.gov

Figure 1. Donald Hughes book library at the National Nuclear Data Center, Brookhaven National Laboratory.

Over the years the data compilation efforts have evolved and spread worldwide which was facilitated by the nuclear data developments in Vienna, Austria (Nuclear Data Section or NDS-IAEA), Paris, France (Nuclear Energy Agency or NEA-Databank) and Obninsk, USSR (Center Jadernykh Dannykh or CJD) in 1963-1964. In 1970 the four original data centers agreed on the data interchange or exchange format (EXFOR). This agreement led to the formation of the Nuclear Reaction Data Centres (NRDC) network under the auspices of the International Atomic Energy Agency (IAEA) to oversee the neutron data compilations, and the NRDC will celebrate its golden jubilee in 2020. Authority for reaction data compilations were assigned on a geographical basis: Area #1: U.S. and Canada (NNDC), Area #2 Western Europe + Japan (NEA-Databank), Area #3 Eastern Europe, Africa, Asia, Australia, Latin America (NDS-IAEA) and Area #4: Former USSR (CJD).

The geographical separation of compilation work has simplified interactions between researchers and data compilers. This joint effort resulted in a successful international collaboration and produced the only continuously-updated low- and intermediate-energy nuclear reaction EXFOR database, and these data are publicly available from the NDS-IAEA (https://www-nds.iaea.org/exfor/) or NNDC (https://www.nndc.bnl.gov/exfor/) web sites. Fig. 2 highlights the distribution of ~18 million data points supplied to EXFOR by 12 active and 3 discontinued centers. The database major contributors include NNDC (Brookhaven, USA), NEA-DB (OECD, Paris, France), NDS (IAEA, Vienna, Austria), CNPD (Sarov, Russian Federation) and JCPRG (Sapporo, Japan). The graphic data show that the majority of complex measurements were performed in the USA and Europe, and NNDC (Area #1) and NEA-Databank (Area #2), NDS-IAEA (Area #3) provide the largest individual contribution to the library. These centers are closely followed by the data efforts in the Russian Federation and Japan.

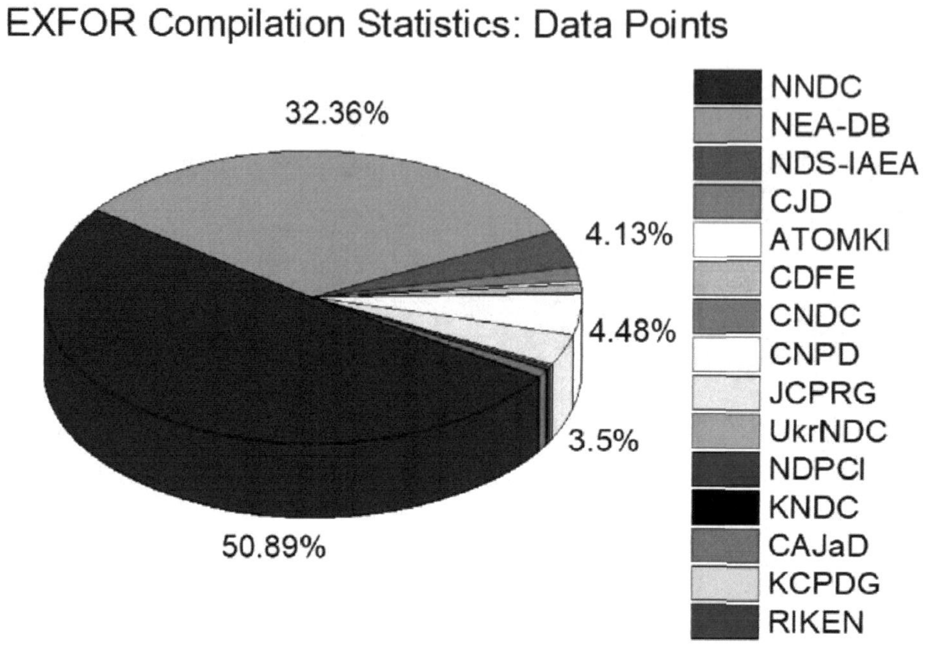

Figure 2. EXFOR compilation statistics include data points from the currently operating NNDC Brookhaven, NEA-DB Paris, NDS IAEA Vienna, CJD Obninsk, ATOMKI Debrecen, CDFE Moscow, CNDC Beijing, CNPD Sarov, JCPRG Sapporo, UkrNDC Kyiv, NDPCI BARC/India, KNDC Taejon and discontinued CAJaD, KCPDG and RIKEN data centers. Complete centers descriptions and complementary details are available from the EXFOR compilation control system: https://www-nds.iaea.org/public/exfor/x4compil/. As of November 11, 2019 the total number of data points was 17,467,585.

2 Missing Data

Today, EXFOR library and its computer database are primary repository of experimental neutron, charged-particle, and photonuclear reaction data compilations. However, the broad scope of compilations was not a case at the beginning, and the library has evolved due to historical and technological reasons. Fig. 3 shows that the initial compilation area was focused only on neutron cross section and spontaneous fission data while many important nuclear physics quantities such as neutron fission yields, charged-particle and photonuclear reaction parameters were not compiled until the compilation scope was changed in 1976. In those days, computer technologies were very immature and required large number of people to produce the relatively small compilations based on tabulated data. The lack of Internet access and general unavailability of laboratory reports [9] produced a situation wherein many important results were unnoticed by the EXFOR compilers. These missing EXFOR references were analyzed in our previous publication [10], and the results are shown graphically in Fig. 4.

To further understand the situation we would consider the case of ^{254}Cf spontaneous fission. Tentative observations of the heavy element production in neutron stars merger (GW010817) [11–13] created plenty of enthusiasm in the nuclear astrophysics community, and several complementary scenarios were developed. One of the most interesting possibilities includes analysis of Kilonova light curves for the possible impact of ^{254}Cf spontaneous

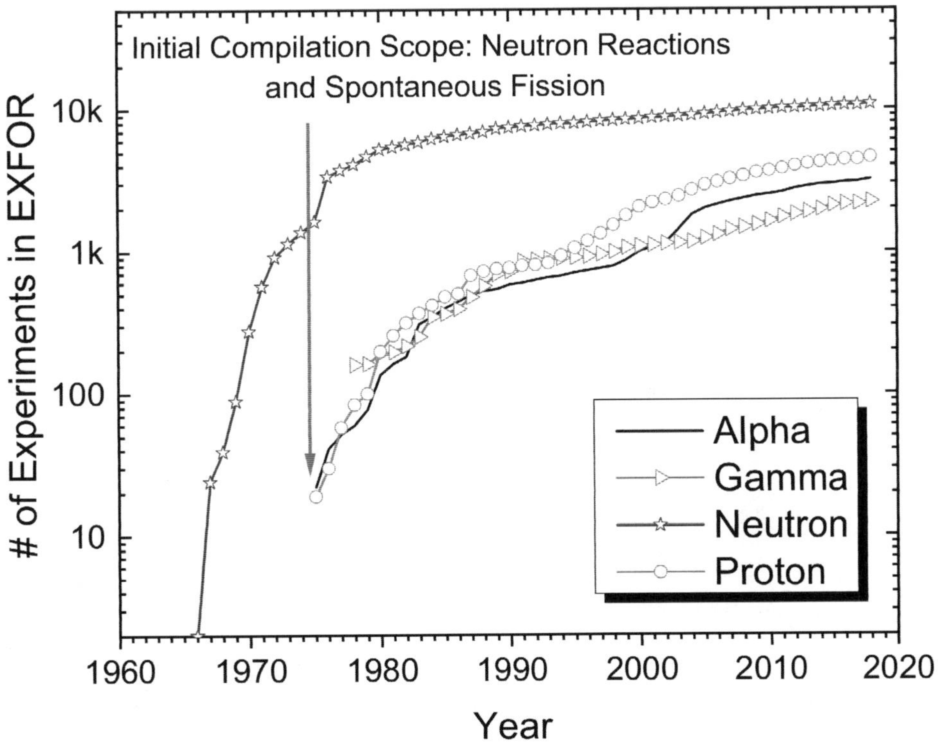

Figure 3. EXFOR compilation timeline shows evolution of the library compilation scope and data contents over the years.

fission [14]. The authors of this paper could not find relevant ^{254}Cf spontaneous fission data using the EXFOR database [6] and resorted to theoretical calculations. However, the decays of the californium isotopes have been extensively studied by J. Wilhelmy's group at Los Alamos National Laboratory. In fact, the substantial amount of ^{254}Cf ($\sim 10^{12}$ nuclei) has been produced at the TRU (trans-uranium) facility at Oak Ridge National Laboratory as part of the trans-uranium production programme of Energy Research and Development Administration (ERDA) for a USA /Israel group. The results were produced at the Weizmann Institute, Rehovot, Israel and published as the Grenoble conference proceedings and a Jülich, Germany laboratory report [15–17]. It was very common in the 70s when scientists published their findings in secondary publications [9] that were not accessible to EXFOR compilers. The recent publication of Kilonova light curves analysis [14] has motivated the nuclear data community to re-examine this situation using complementary resources. The analysis of the Nuclear Science References database [18] by searching for ^{254}Cf experimental data reveals several experiments [16, 17] in which spontaneous fission of ^{254}Cf was extensively studied. These previously missed fission yields were presented to the astrophysics community and experimental data compilations are in progress.

3 Analysis of Fission Yields

It is common knowledge that any new measurements are very expensive. For instance, the cost of irradiation of electronic components for space flights at the 88" Lawrence Berkeley

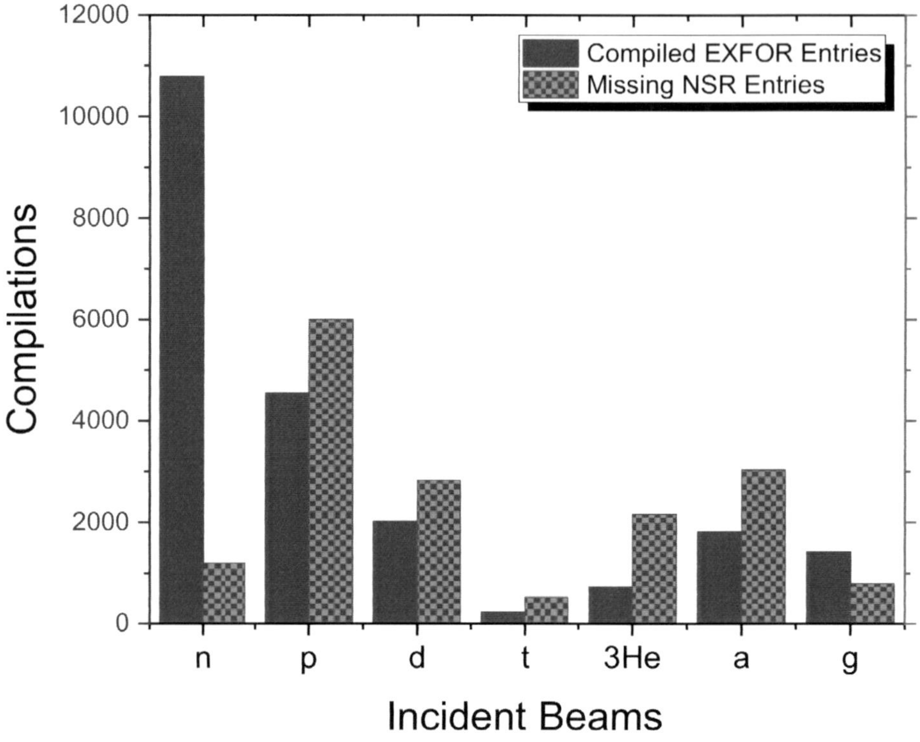

Figure 4. EXFOR completeness for neutron-, charged- and photonuclear-induced reactions.

National Laboratory is \$ 2,500 per hour [19]. This results in the \$ 420 K beam time price tag for a typical one-week experiment. Addition of the target, electronics and staff expenses brings the overall cost to \$ 1 M. Therefore, it is very cost effective to recover the previously-published results before conducting a new measurement. Further analysis allows us to conservatively estimate the monetary value of ~23,000 EXFOR experiments is approximately \$ 23 B. This finding clearly demonstrates the importance of the present work and the strong needs for nuclear reaction data preservation, storage and dissemination.

Recently, the overall situation with missing in EXFOR fission yields compilations has been analyzed at the NNDC using the NSR database that is based on the Donald Hughes book library [10]. These results were summarized in three NRDC memos [20–23] that provided a guide for the network to fix the EXFOR database deficiencies. A similar effort is currently underway at the Nuclear Data Section, IAEA, Vienna [24]. The IAEA group has analyzed references from the U.S. and U.K. fission yields evaluations [25, 26]. This complementary effort would ensure the completeness of the EXFOR database for fission data. In addition, the fission yield evaluation references [25, 26] were explored at NNDC for missing in NSR articles, PDF files were collected, and publications were entered into the database. These developments guarantee the mutual completeness of the NSR and EXFOR databases that is needed for nuclear reaction and structure evaluations.

The missing data analysis has been used to estimate the workload for the Area #1 fission yields compilations that are now in progress. By the end of Fiscal Year 2019, 18 new neutron-, 6 new photo-induced fission and 41 revised neutron-induced fission compilations have been submitted to the NRDC network. The compilation of previously missed photofission work of Smith and Richardson [27] is shown in Fig. 5. The new reaction entry consists of one bibliographical and two data subentries. The first data subentry contains independent fission yields for isotopes of Br, Rb, Nb, Sb and Cs. The second data subentry contains chain fission yields for A=125,127. These data have been processed in EXFOR format using the Sarov EXFOR editor [28]. The NNDC compilation project is currently going through the Brookhaven group memos [20–23]. Work on the IAEA group findings [24] should start in 2020, and the tentative target completion is Fall 2020.

4 Conclusion

The current work reviews the present status of fission yields experimental data and describes compilation work in Area #1. The missing database references have been identified and corresponding PDF files for the articles were recovered. These materials were shared with the NRDC network, and the compilation activities are currently underway at the NNDC. The examples of the NSR database queries for missing references, data analyses, and compilations are given. NSR and EXFOR databases comparison demonstrated the large potential of both databases and produced many complementary findings. The results of this work are an integral part of the BNL-Los Alamos reevaluation of neutron fission yields [29] that would make a broad impact in multiple areas of applied and fundamental science.

The authors are indebted to A. Sonzogni (BNL) for encouragement of this project, T. Kawano (LANL) for useful comments and R. Arcilla (BNL) for a careful reading of the manuscript and valuable suggestions. We are grateful to the NRDC network members for the continuous work on EXFOR data compilations and overall database improvements. We acknowledge partial support from the NA-22 grant number 000024158. Work at Brookhaven was funded by the Office of Nuclear Physics, Office of Science of the U.S. Department of Energy, under Contract No. DE-AC02-98CH10886 with Brookhaven Science Associates, LLC.

References

[1] D.J. Hughes, T.W. Bonner, H. Goldstein, W.W. Havens, Jr., I. Kaplan, C.O. Muehlhause, A.H.Snell, J.R. Stehn, T.M. Shyder, R.F. Taschek, A. Wattenberg, "A Compilation of the AEC Neutron Cross Section Advisory Group," Brookhaven National Laboratory Report 170, May 15 (1952).

[2] H.H. Goldsmith, W.W. Ibser, "Neutron Cross Sections of the Elements," United States Atomic Energy Commission Report, MMDC 59 (1946).

[3] H.H. Goldsmith, W.W. Ibser, B.T. Feld, "Neutron Cross Sections of the Elements A Compilation," Review of Modern Physics **19**, 259 (1947).

[4] S.F. Mughabghab, Atlas of Neutron Resonances, Resonance Properties and Thermal Cross Sections Z=1-60 **1**, Elsevier Publisher, Amsterdam (2018).

[5] D.A. Brown, M.B. Chadwick, R. Capote, A.C. Kahler, A.Trkov, M.W. Herman, A.A. Sonzogni, Y. Danon, A.D. Carlson, M. Dunn, D.L. Smith, G.M. Hale, G. Arbanas, R. Arcilla, C.R. Bates, B. Beck, B. Becker, F. Brown, R.J. Casperson, J. Conlin, D.E. Cullen, M.-A. Descalle, R. Firestone, T. Gaines, K.H. Guber, A.I. Hawari, J. Holmes, T.D. Johnson, T. Kawano, B.C. Kiedrowski, A.J. Koning, S. Kopecky, L. Leal, J.P. Lestone, C. Lubitz, J. I.Marquez Damian, C.M. Mattoon, E.A. McCutchan, S. Mughabghab, P. Navratil, D. Neudecker, G.P.A. Nobre, G. Noguere, M.Paris, M.T. Pigni, A.J. Plompen, B. Pritychenko, V.G. Pronyaev, D. Roubtsov, D. Rochman, P.Romano, P. Schillebeeckx, S. Simakov, M. Sin, I. Sirakov, B. Sleaford, V. Sobes, E.S. Soukhovitskii, I. Stetcu, P. Talou, I. Thompson,

S. van der Marck, L. Welser-Sherrill, D. Wiarda, M. White, J. L.Wormald, R. Q.Wright, M. Zerkle, G. Zerovnik, Y. Zhu, Nucl. Data Sheets **148**, 1 (2018).

[6] V.V. Zerkin, B. Pritychenko, "The experimental nuclear reaction data (EXFOR): Extended computer database and Web retrieval system," Nucl. Instrum. Meth. Phys. Res. A **888**, 31 (2018).

[7] N. Otuka, E. Dupont, V. Semkova, B. Pritychenko, A.I. Blokhin, M. Aikawa, S. Babykina, M. Bossant, G. Chen, S. Dunaeva, R.A. Forrest, T. Fukahori, N. Furutachi, S. Ganesan, Z. Ge, O.O. Gritzay, M. Herman, S. Hlavac, K. Kato, B. Lalremruata, Y.O. Lee, A. Makinaga, K. Matsumoto, M. Mikhaylyukova, G. Pikulina, V.G. Pronyaev, A. Saxena, O. Schwerer, S.P. Simakov, N. Soppera, R. Suzuki, S. Takacs, X. Tao, S. Taova, F. Tarkanyi, V.V. Varlamov, J. Wang, S.C. Yang, V. Zerkin, Y. Zhuang,"Towards a More Complete and Accurate Experimental Nuclear Reaction Data Library (EXFOR): International Collaboration between Nuclear Reaction Data Centres (NRDC)," Nucl. Data Sheets **120**, 272 (2014).

[8] N. Holden, "A Short History of CSISRS: At the Cutting Edge of Nuclear Data Information Storage and Retrieval Systems and its Relationship to CINDA, EXFOR and ENDF," December 2005. Downloaded from ⟨https://www.nndc.bnl.gov/exfor/compilations/CSISRSHistory.pdf⟩ on June 26, 2019.

[9] B. Pritychenko, "Evolving landscape of low-energy nuclear physics publications," Scientometrics **109**, 2067 (2016).

[10] B. Pritychenko, O.Schwerer, J. Totans, V. Zerkin, O. Gritzay, "Completeness of Neutron-, Photo-induced and Spontaneous Fission Yields Data," Eur. Phys. J. Cof. (2020).

[11] E. Pian, P. D'Avanzo, S. Benetti, *et al.*, Nature(London) **551**, 67 (2017).

[12] S.J. Smartt, T.-W. Chen, A. Jerkstrand, *et al.*, Nature(London) **551**, 75 (2017).

[13] N.R. Tanvir, A.J. Levan, C. Gonzalez-Fernandez, *et al.*, Astrophys. J. **848**, L27 (2017).

[14] Y. Zhu, R.T. Wollaeger, N. Vassh, R. Surman, T.M. Sprouse, M.R. Mumpower, P. Möller, G.C. McLaughlin, O. Korobkin, T. Kawano, P.J. Jaffke, E.M. Holmbeck, C.L. Fryer, W.P. Even, A.J. Couture, J. Barnes, "Californium-254 and Kilonova Light Curves," Astrophys.J. **863**, L23 (2018).

[15] E. Cheifetz, H.A. Selic, A. Wolf, J.B. Wilhelmy, "Even-Even Neutron-Rich Isotopes," Contrib. Conf. Nucl. Spectrosc. Fission Products, Grenoble, 193 (1979).

[16] H.A. Selic, E. Cheifetz, A. Wolf, J.B. Wilhelmy, "Neutron-Rich Fragments from Spontaneous Fission of ^{254}Cf," Contrib. Conf. Nucl. Spectrosc. Fission Products, Grenoble, 316 (1979).

[17] H.A. Selic, E. Cheifetz, J.B. Wilhelmy, "Studies of Primary Fission Products from ^{254}Cf," JUL-Spez-99, 69 (1981).

[18] B. Pritychenko, E. Betak, M.A. Kellett, B. Singh, J. Totans, "The Nuclear Science References (NSR) database and Web Retrieval System," Nucl. Instr. Meth. Phys. Res. A **640**, 213 (2011).

[19] L. Bernstein, private communication to B. Pritychenko, March (2019).

[20] O. Schwerer, B. Pritychenko, "Completeness check EXFOR vs. NSR: Spontaneous Fission Yields," NRDC Memo CP-C/464, NDS-IAEA, 29 November 2018.

[21] O. Schwerer, B. Pritychenko, "Completeness check EXFOR vs. NSR: Photofission Yields," NRDC Memo CP-C/465, NDS-IAEA, 30 January 2019.

[22] O. Schwerer, B. Pritychenko, "Completeness check EXFOR vs. NSR: Neutron-induced fission yields," NRDC Memo CP-C/466, NDS-IAEA, 19 February 2019.

[23] O. Schwerer, "Procedure used for the completeness check EXFOR vs. NSR on fission yields," NRDC Memo CP-C/469, NDS-IAEA, 14 April 2019.

[24] S. Okumura, N. Otuka, et al., Eur. Phys. J. Cof. (2020).

[25] T.R. England, B.F. Rider, "Evaluation and Compilation of Fission Product Yields 1993," Los Alamos National Laboratory Report LA-UR-94-3106 (1994).

[26] R.W. Mills, "Fission product yield evaluation," Thesis, Univ. of Birmingham (1995).

[27] J.R. Smith, A.E. Richardson, "Independent yields from the photofission of 232Th and the Z(p) and statistical-dynamic models," Phys.Rev. C **44**, 1118 (1991).

[28] G.N. Pikulina, S.M. Taova, "Activities of the RFNC-VNIIEF Center of Nuclear Physics Data on the Compilation of Experimental Data for the EXFOR International Library: EXFOR-EDITOR Software Package," Phys.Atomic Nuclei **81**, 1450 (2018).

[29] "Nuclear Data Interagency Working Group / Research Program," DOE National Laboratory Announcement Number: LAB 18-1903. Downloaded from ⟨https://www.energy.gov/science/np/nuclear-physics⟩ on June 26, 2019.

```
ENTRY            L0253   20190523                                  L0253  0     1
SUBENT        L0253001   20190523                                  L0253  1     1
BIB                 11         25                                  L0253  1     2
TITLE      Independent yields from the photofission of 232Th,      L0253  1     3
           and the Z(p) and statistical-dynamic models             L0253  1     4
AUTHOR     (J.R.Smith,A.E.Richardson)                              L0253  1     5
INSTITUTE  (1USANMS)                                               L0253  1     6
REFERENCE  (J,PR/C,44,1118,1991)                                   L0253  1     7
           #doi:10.1103/PhysRevC.44.1118                           L0253  1     8
           #NSR: 1991SM04                                          L0253  1     9
FACILITY   (LINAC,1USAWSA)                                         L0253  1    10
INC-SOURCE (BRST) Three irradiations were performed with 27-MeV    L0253  1    11
           peak bremsstrahlung, one with 15 MeV, one with 10 MeV.  L0253  1    12
           Fluctuations in peak bremsstrahlung energies during theL0253  1    13
           irradiations were about +-1 MeV.                        L0253  1    14
METHOD     (ACTIV) The duration of each irradiation was            L0253  1    15
           approximately 1 h.                                      L0253  1    16
           (CHSEP,GSPEC)                                           L0253  1    17
DETECTOR   (GELI,GE-IN) A Canberra Ge(Li) detector with a 2.3-MeV  L0253  1    18
           resolution for the 1.33-MeV gamma ray from Co and a 16%L0253  1    19
           efficiency, two Ortec Ge(Li) detectors with a 2.1-keV   L0253  1    20
           resolution and a 15% efficiency, and an Ortec intrinsicL0253  1    21
           Ge detector with a 1.75-keV resolution and 25%          L0253  1    22
           efficiency.                                             L0253  1    23
SAMPLE     40-g samples of reagent-grade Th(NO3)4*4H20 in          L0253  1    24
           polyethylene ampoules                                   L0253  1    25
ERR-ANALYS (DATA-ERR) No information on sources of uncertainties   L0253  1    26
HISTORY    (20190523C) OG.                                         L0253  1    27
ENDBIB              25          0                                  L0253  1    28
NOCOMMON             0          0                                  L0253  1    29
ENDSUBENT           28          0                                  L0253  199999
SUBENT        L0253002   20190523                                  L0253  2     1
BIB                  2          2                                  L0253  2     2
REACTION   (90-TH-232(G,F)ELEM/MASS,IND,FY,,BRA)                   L0253  2     3
STATUS     (TABLE) Table I from Phys.Rev.,C44,1118,1991            L0253  2     4
ENDBIB               2          0                                  L0253  2     5
NOCOMMON             0          0                                  L0253  2     6
DATA                 6          9                                  L0253  2     7
EN-MAX     MASS       ELEMENT    DATA-MAX   DATA       DATA-ERR    L0253  2     8
MEV        NO-DIM     NO-DIM     PC/FIS     PC/FIS     PC/FIS      L0253  2     9
   11.        136.        55.                  1.06E-3    0.48E-3L0253  2    10
   15.        136.        55.                  3.37E-3    0.36E-3L0253  2    11
   27.         82.        35.                  8.91E-5    1.18E-5L0253  2    12
   27.         86.        37.        5.84E-4                     L0253  2    13
   27.         96.        41.                  2.40E-5    0.36E-5L0253  2    14
   27.        124.        51.                  1.26E-3    0.15E-3L0253  2    15
   27.        126.        51.                  2.41E-2    0.25E-2L0253  2    16
   27.        134.        55.        1.42E-4                     L0253  2    17
   27.        136.        55.                  1.13E-2    0.12E-2L0253  2    18
ENDDATA             11          0                                  L0253  2    19
ENDSUBENT           18          0                                  L0253  299999
SUBENT        L0253003   20190523                                  L0253  3     1
BIB                  3          5                                  L0253  3     2
REACTION   (90-TH-232(G,F)MASS,CHN,FY,,BRA)                        L0253  3     3
ANALYSIS   The mass yields for masses 125 and 127 were based on    L0253  3     4
           the assumption that the cumulative chain yields of      L0253  3     5
           125Sb and 127Sb represented the mass yields.            L0253  3     6
STATUS     (TABLE) Table I from Phys.Rev.,C44,1118,1991            L0253  3     7
ENDBIB               5          0                                  L0253  3     8
COMMON               1          3                                  L0253  3     9
EN-MAX                                                             L0253  3    10
MEV                                                                L0253  3    11
   27.                                                             L0253  3    12
ENDCOMMON            3          0                                  L0253  3    13
DATA                 3          2                                  L0253  3    14
MASS       DATA       DATA-ERR                                     L0253  3    15
NO-DIM     PC/FIS     PC/FIS                                       L0253  3    16
   125.       0.29       0.06                                      L0253  3    17
   127.       0.44       0.05                                      L0253  3    18
ENDDATA              4          0                                  L0253  3    19
ENDSUBENT           18          0                                  L0253  399999
ENDENTRY             3          0                                  L025399999999
```

Figure 5. EXFOR compilation of ^{232}Th(γ,F) work of J.R. Smith and A.E. Richardson [27].

Completeness Check of Experimental Fission Product Yield Data in EXFOR Database

Shin Okumura[1,*]

[1]NAPC-Nuclear Data Section, International Atomic Energy Agency

Abstract. The IAEA conducted the completeness assessment of fission product yield (FPY) data in EXFOR database against two bibliographies that were used in ENDF/B-VI and UKFY3.0 library evaluations. We found that 194 articles are found to be relevant for new entry.

1 Introduction

Since the discovery of nuclear fission, the fission product yield (FPY) is of interest from the fundamental understandings of nuclear physics as well as many nuclear energy applications. The nuclear fission is a complex phenomenon, whereby the nucleus forms compound, distort the shape, splits into two smaller fragments and release a large amount of energy with large amount of energy release. Until now, a complete picture of nuclear fission is not well understanded by any theories. It owing to the fact that the difficulty of sample preparation and experimental handling of heavy actinides.

The evaluation of major nuclear data libraries has been done mostly in 1990s. They generally cover fission yield data from the neutron-induced and spontaneous fission of 20–30 fissile materials including ones without sufficient experimental data. Therefore, such evaluation has been accomplished with both a guidance of available experimental data and utilizing empirical models [1, 2] to have consistencies in other fission observable such as prompt fission neutron multiplicity.

From 1990s, the IAEA Nuclear Data Section (NDS) has widely reviewed the user needs in these application fields in the previous IAEA Coordinated Research Projects (CRPs), *e.g.*, "Compilation and evaluation of fission yield nuclear data" (1991-1996) [3], "Fission Product Yield Data for the Transmutation of Minor Actinide Nuclear Waste" (1997- 2002) [4], and the Technical Meeting on "Fission Product Yields: current status and perspectives" (2016) [5] to focus on the current situations of the experimental, modeling, and theoretical points of view for the FPY data evaluation. The recent increase in demand of the FPY data in various fields has triggered activities of developing new methods, codes and models for FPY evaluation. The IAEA-NDS will launch a nre CRP in 2020 for updating the evaluated FPY data for the neutron-induced fission of some important actinides.

The reliability of the evaluated data depends on the quality and authenticity of experimental data used in evaluation, where a complete set of the experimental FPY data is crucial. However, the experimental data stored in the international nuclear reaction database, also known as EXFOR [6], does not entirely cover all available FPY data. In this report, we

*e-mail: s.okumura@iaea.org

outlines the procedures of completeness check of FPY data in the EXFOR database by comparing with the bibliographical references cited in the summary reports of the ENDF-B/VI [7] and UKFY3.0 [8] evaluations.

2 Completeness assessment

Since the procedure of completeness assessment of FPY data in EXFOR database comparing with FPY data used in ENDF-B/VI [7] and UKFY3.0 [8] evaluation summaries has been describe elsewhere [9–12], a brief description is given here.

The bibliographies of the ENDF-B/VI [7] and UKFY3.0 [8] evaluation summaries are converted into `Author` and `Reference` format that defined in the EXFOR format manual [13] (*e.g.*,`J,PR,60,534,1941` for Phys. Rev. 60(1941)534). Two lists are merged and stored into the SQL database. `Reference` data are double-checked manually followed by a brief check using Python script, then `Reference` data are used in search of records EXFOR and CINDA. In case of a particular `Reference` has not found neither in EXFOR nor in CINDA, we identify the article using Crossref database for journal articles, online libraries, search engines, or electronic repositories such as International Nuclear Information System (INIS). If the electronic format is not available, hard copy of the article are collected through the IAEA Library.

The ENDF/B-VI evaluation summary [7] cites 1,602 articles in total, whose 924 articles have used to the actual evaluations while the rest of 678 articles have not used. The UKFY3.0 [8] evaluations cites 645 articles. Table 1 shows the number of articles categorised during our completeness assessment. Approximately 23% of articles are overlapped in two bibliographies. After exclusion of the overlapped articles, the number of articles found to to be newly compiled are 194, while the articles that need to be checked with existing EXFOR entries are 54. The compilation has been started from April 2019 by NRDC data centres.

Figure 1 shows the number of FPY entries in EXFOR, which reflects the number of publications in a year. The first FPY measurements in EXFOR database was published in 1943. Until 1960s, FPY measurements are mostly dominated by the neutron induced fission of important actinides as a targets, namely Th, U, Pu, and Cf isotopes. The experimental FPY data drastically increase in 1970s. After 1980s, variations of target isotopes, *e.g.* Np, Am or Cm, have increased while the majority of trend remains. Up to now, the majority of reactions are the neutron induced fission followed by the the spontaneous fission, photo-nuclear, and proton induced fission. By comparing number of FPY data in Fig. 1 extracted from EXFOR in July 2019 and December 2019, the number of entries increase in wide range of published year by efforts of NRDC data centres.

Table 1. The number of FPY references categorised during our assessment for those cited in evaluation summaries of ENDF/B-VI [7] and UKFY3.0 [8]. See Refs. [9–12] for details.

Source	Exist[a]	Create new entry[b]	Check[c]	Not relevant[d]	Total
ENDF/B-VI	618	154	36	116	924
UKFY3.0	409	139	32	64	645
Total[e]	720	194	54	167	1203

[a] Already compiled in EXFOR.
[b] Not found in EXFOR and need to create new entry.
[c] Need to check if all relevant data are compiled in EXFOR entry.
[d] Not FPY or not relevant for EXFOR or the original article cannot identify.
[e] Total after elimination of overlap.

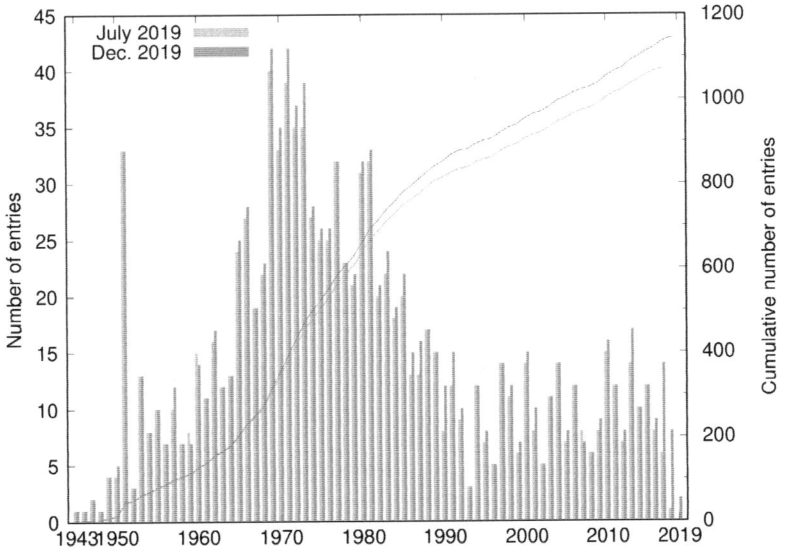

Figure 1. Number of FPY data entries in EXFOR based on the published year.

3 Summary

We reviewed the completeness of the FPY data in the EXFOR database by crosschecking against the bibliographies of the ENDF-B/VI and UKFY3.0 evaluation summaries. It was found that 194 references are relevant to new compilation of EXFOR entry, and 54 references must be checked with the existing EXFOR entries to ensure that the all relevant FPY data are properly compiled. The compilation has been started by data centres and the number of entries clearly increased after a half year from it has started.

References

[1] A.C. Wahl, Tech. Rep. LA-13928, Los Alamos National Laboratory (2002)
[2] D.G. Madland, T.R. England, Tech. Rep. LA-6595-MS, Los Alamos National Laboratory (1994)
[3] IAEA, Tech. Rep. IAEA-TECDOC-1168, IAEA (2000)
[4] IAEA, Tech. Rep. STI/PUB/1286, IAEA (2008)
[5] P. Dimitriou, F.J. Hambsch, S. Pomp (eds.), Tech. Rep. INDC(NDS)-0713, IAEA (2016)
[6] *Experimental Nuclear Reaction Data (EXFOR)*, https://www-nds.iaea.org/exfor/
[7] T.R. England, B. Rider, Tech. Rep. ENDF-349, LA-UR-94-3106, Los Alamos National Laboratory (1994)
[8] R. Mills, Ph.D. thesis, University of Birmingham (1995)
[9] S. Okumura, NRDC Meeting 2019 Working paper (2019)
[10] S. Okumura, NRDC Memo CP-D/979 (2019)
[11] M. Fleming, T. Kawano, N. Otuka (eds.), Tech. Rep. INDC(NDS)-793, International Atomic Energy Agency (2019)
[12] T. Fukuda, S. Okumura, N. Otuka, submitted for EPJ Web of Conferences (2020)
[13] O. Schwerer (ed.), Tech. Rep. IAEA-NDS-207 Rev.2015/08, IAEA (2015)

Microscopic Calculation of Fission Fragment Distributions in Actinides

N. Schunck[1]*

[1]Nuclear and Chemical Sciences Division, Lawrence Livermore National Laboratory, Livermore, CA 94551, USA

Abstract. The simulation of independent and cumulative yields requires precise knowledge of the initial conditions of the fission fragments immediately after scission. In this paper, we use a quantum-mechanical description of fission dynamics to extract the initial mass distribution of fission fragments for the neutron-induced fission of the two major actinides ^{239}Pu and ^{235}U, both for thermal fission and as a function of the incident neutron energy.

1 Introduction

Nuclear fission remains a fascinating basic science problem, which challenges our understanding of both nuclear forces and the structure of interacting quantum many-body fermion systems. A lot of progress has been made to build a microscopic description of this process, where fission observables would be computed within a fully quantum-mechanical theory [1]. Nuclear density functional theory (DFT) is currently the most natural framework to achieve this, as it is naturally tailored to handle heavy nuclei for which most experimental data on fission is available [2]. As we progress in the path to building a comprehensive, predictive theory of fission, it is important to establish a number of reference points along the way.

The goal of this paper is precisely to report such benchmark calculations. In this paper, we are going to focus on the initial fission fragment mass distributions (before prompt emission) for the particular cases of ^{239}Pu(n,f) and ^{235}U(n,f). In addition to results for thermal fission, we are also going to show results as a function of the incident energy of the neutron. In Section 2, we briefly recall the theoretical framework that we used while results are presented in Section 3.

2 Theoretical Framework

The charge and mass distributions of fission fragments are extracted by explicit simulation of fission dynamics within a fully quantum-mechanical framework. Specifically, we work within the time-dependent generator coordinate method (TDGCM) under the Gaussian overlap approximation (GOA). As discussed in details in [3–10], the GOA allows recasting the Hill-Wheeler equation of the TDGCM into a collective, Schrödinger-like equation of motion for a collective wave package $g(\boldsymbol{q}, t)$ in the multi-dimensional collective space C spanned by collective variables $\boldsymbol{q}$. In practice, we used the code FELIX-2 to solve the collective Schrödinger

*e-mail: schunck1@llnl.gov

equation and extract the fission fragment distributions $Y(A)$ from the flux of the wave packet through the scission line; for details about the general methodology, see [9].

The collective space C is generated by solving the HFB equation with constraints on the expectation value of the axial quadupole and axial octupole moments, $\boldsymbol{q} \equiv (q_{20}, q_{30})$. All HFB calculations have been performed with the SkM* parametrization [11] of the Skyrme energy density functional. The pairing channel is described with a surface-volume, density-dependent pairing force. The strengths $V_{0,n}$ and $V_{0,p}$ of this interaction were fitted in each nucleus of interest in order that the HFB pairing gap in the ground-state reproduce the experimental value of the 3-point odd-even staggering indicator [12]. For ^{240}Pu, this gives: $V_{0,n} = -265.250$ MeV and $V_{0,p} = -340.062$ MeV, and for ^{236}U: $V_{0,n} = -255.250$ MeV and $V_{0,p} = -325.594$ MeV. The HFB equation is solved with the HFODD solver by expanding the solutions on the one-center harmonic oscillator (HO) basis [13]. We used the exact same characteristics as in [14].

The practical implementation of the TDGCM+GOA also requires the knowledge of the collective inertia, or mass, tensor $\mathsf{B}(\boldsymbol{q})$, the metric tensor $\gamma(\boldsymbol{q})$, and the zero-point energy correction $\varepsilon_{\mathrm{ZPE}}(\boldsymbol{q})$. All these quantities were computed from the GCM formula at the perturbative cranking approximation; see Chap. 6 in [2] for details. This choice was made in order to guarantee consistency between all the ingredients of the model.

3 Benchmark Results

We focus on the neutron-induced fission of the two major actinides, ^{239}Pu and ^{235}U, which both lead to an even-even compound nucleus. For each of these two nuclei, we compute the potential energy surface (PES) in the two-dimensional collective space C spanned by collective variables $\boldsymbol{q} \equiv (q_{20}, q_{30})$, where $q_{\lambda\mu}$ is the expectation value of the mass multipole moment of multipolarity λ, μ on the HFB wavefunction. An example of the potential energy surface for ^{236}U is shown in Fig.1; the results for ^{240}Pu are shown in [14]).

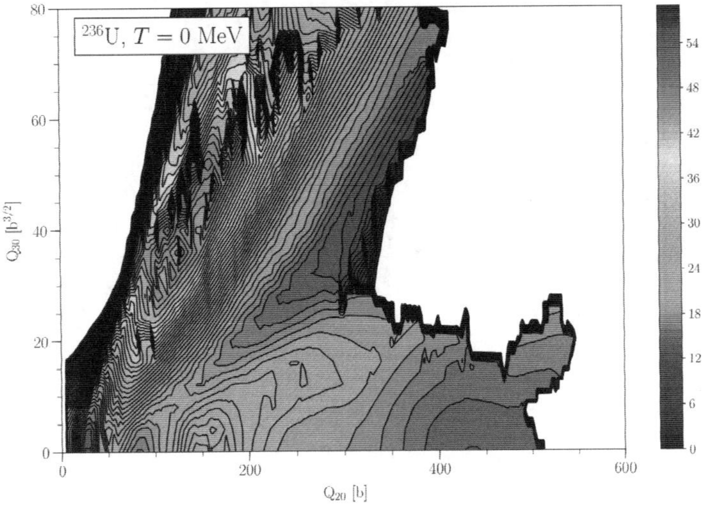

Figure 1. Two-dimensional potential energy surface for ^{236}U. The energy is computed at the HFB level with the SkM* Skyrme functional and the pairing force as described in the text.

In this benchmark, we define scission configurations by the condition that the expectation value of the Gaussian neck operator is lower than q_{cut}. Since there is a considerable amount of ambiguity on the actual definition of scission – see discussion in [1, 14, 15] and references therein, we look at three possible choices $q_{\text{cut}} = 4, 5$ and 6. Given this definition, the PES and the collective mass tensor associated with each point of said PES, we run the code FELIX and solve the collective Schrödinger-like equation for collective dynamics. To guarantee convergence of the results, we adopt a finite-element grid characterized by $h = 6$ and run the solver up to $t_{\text{final}} = 25$ zs [10, 16]. The initial state is constructed as a Gaussian superposition of GCM states such the average energy is about 1 MeV above the saddle point.

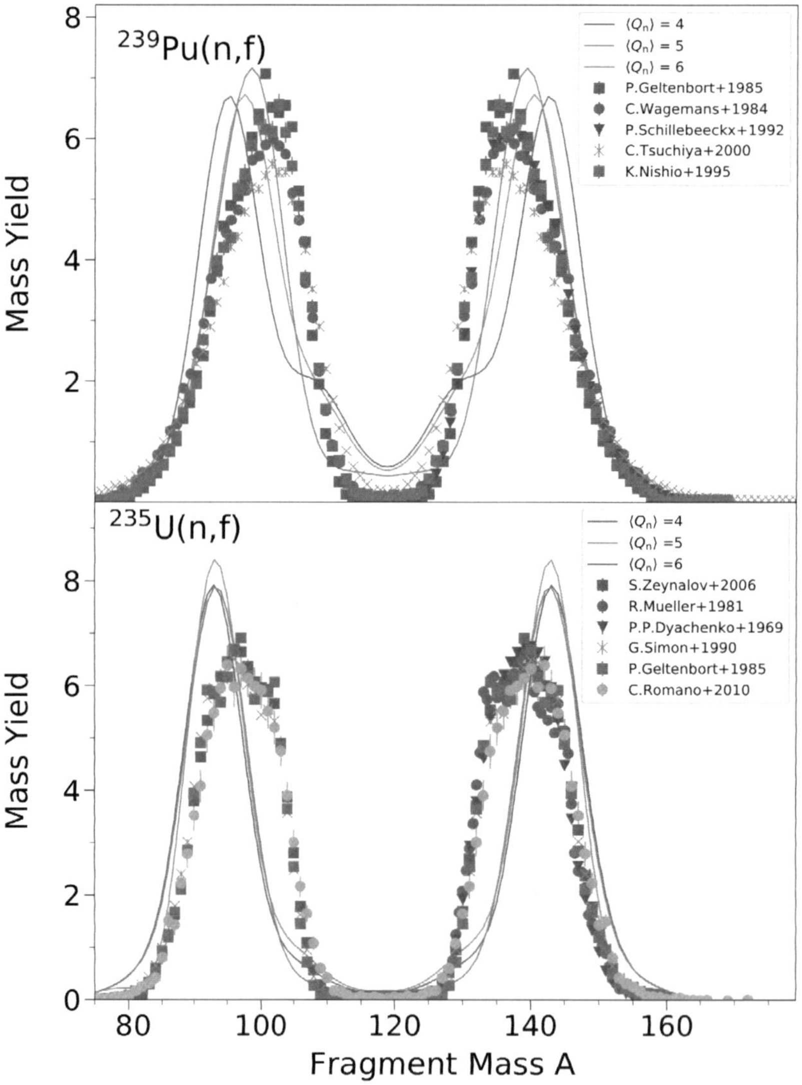

Figure 2. Initial fission fragment mass distributions for the ^{239}Pu(n,f) (top) and ^{235}U(n,f) (bottom) thermal fission. Calculations for three different different definitions of the scission line are compared with existing data from the EXFOR database.

Figure 2 shows the resulting mass distributions for both actinide nuclei. The results for ^{239}Pu are rather good and do not show too much variation with the definition of scission. The case of ^{235}U is not nearly as good as the yields are too narrow and too widely separated. This suggests that the collective space is too small: since mass yields are normalized, fragmentations missing from the scission line lead to artificially higher yields elsewhere. This problem was already identified in [17], where it was suggested that (q_{20}, q_{30}) should be replaced by collective variables more closely aligned with the number of particles in the fragments.

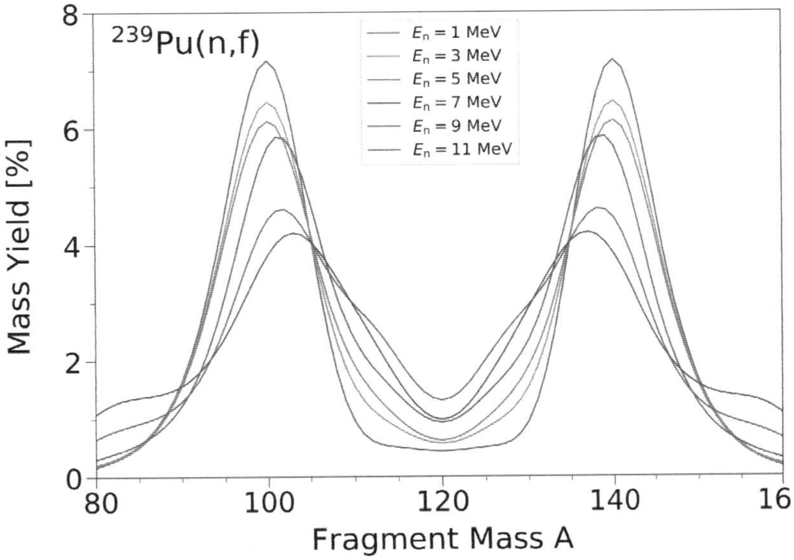

Figure 3. Initial fission fragment mass distributions for the ^{239}Pu(n,f) fission as a function of the incident neutron energy.

In the TDGCM+GOA, one can simulate increasing neutron energies E_n by simply increasing the collective energy, that is, the expectation value of the TDGCM+GOA collective Hamiltonian, $E_n \approx E_{\text{col}} - E_{\text{saddle}}$, with E_{saddle} the HFB energy of the saddle point. Figure 3 shows the evolution of the initial yields as a function of the neutron energy for ^{239}Pu(n,f). These results are qualitatively similar to earlier calculations by Younes and Gogny [17] and compatible with expectations: as the available excitation energy increases, the distribution becomes more symmetric. Although this result is encouraging, it is important to bear in mind that it contains a number of important approximations, which we discuss in more details in the next section.

4 Discussion

Because of the complexity of the fission process, calculations (whether microscopic or not) have a rather large uncertainty budget. In this section, we wish to discuss briefly some of the shortcomings of the particular calculations reported here, especially since there has been progress in addressing some of them.

There is a large consensus in the theory community that the PES is perhaps the most important ingredient in fission calculations [18]. Whether the initial yields are symmetric or asymmetric can be simply determined by whether the least-energy fission pathway goes through reflection-asymmetric shapes (in our calculations, characterized by $Q_{30} \neq 0$). However, these PES clearlt depend on the underlying energy functional. There has recently been much progress in rigorously quantifying the uncertainties associated with the functional [19–21]. Obviously, the PES also depends on the choice of the collective degrees of freedom. While the PES of Fig. 1 only involves collective coordinates, its physics content could be improved, e.g. by including pairing degrees of freedom [22–25]. Progress in computing facilities, combined with publicly available and carefully benchmakred DFT solvers [13, 26] provide more options to solve the HFB equations across an entire PES and handle the issue of its discontinuities [27].

Given the PES, the calculation of the initial yields within the TDGCM+GOA requires in addition full specification of (i) and collective inertia tensor, (ii) scission configurations and (iii) fission fragment properties.

- As discussed early on, scission is not a well-defined concept and there exist several possible definitions [14, 15, 28, 29]. Here we chose a simple criterion based on the size of the neck: as shown in Fig.2, it has a measurable impact on the yield, which could be quantified more rigorously.
- The theory of collective inertia has a long history; see, e.g. [30–34] for a selection of historical review papers on the subject. Recent benchmarks showed that fission observables such as spontaneous fission half-lives are highly sensitive to the approximation used to calculate the tensor B [35].
- Fission fragment properties are in principle determined at scission. In "adiabatic" theories such as the current implementation of the TDGCM+GOA, these properties are not always well predicted. Just before scission, the two prefragments are strongly entangled, leading to some arbitrariness in determining the excitation energy of the fragments [14, 15]; conversely, just after scission, the fragments are cold, contrary to experimental evidence. Going beyond these limitations requires explicit treatment of non-adiabatic effects, which have recently shown great potential [25, 36–40]. In the particular case of particle number, recent applications of projection techniques show great promise to improve the precision of charge and mass yields [41].

Extending the TDGCM+GOA at increasing excitation energy of the compound nucleus opens another Pandora's box of interesting problems. First of all, it is not at all clear what is the best approach to describe the system at excitation energies of the order of 10-20 MeV, where the level density if extremely high: direct approaches where generator states include quasi-particle excitations are still in their infancy and would need to be severely truncated [42]; a finite-temperature formalism poses conceptual problems, both in terms of relating the temperature to the excitation energy [43], and for defining a collective wave function [44]. In the calculations shown in Fig.3, these effects were not taken into account: both the PES and the collective inertia were computed at zero-temperature.

5 Conclusion

In this paper, we report some benchmark calculations of the initial mass yields in the neutron-induced fission of the two major actinides ^{239}Pu and ^{235}U. We compare yields for thermal neutrons with experimental data and show the evolution of the yields as a function of neutron energy. Finally, we discuss some of the theoretical challenges in improving the fidelity of the models.

This work was performed under the auspices of the U.S. Department of Energy by Lawrence Livermore National Laboratory under Contract DE-AC52-07NA27344.

References

[1] N. Schunck, L.M. Robledo, Rep. Prog. Phys. **79**, 116301 (2016)

[2] N. Schunck, *Energy Density Functional Methods for Atomic Nuclei.*, IOP Expanding Physics (IOP Publishing, Bristol, UK, 2019), ISBN 978-0-7503-1423-7, oCLC: 1034572493

[3] J.F. Berger, in *Dynamics of Wave Packets in Molecular and Nuclear Physics*, edited by J. Broeckhove, L. Lathouwers, P. van Leuven (Springer Berlin Heidelberg, 1986), Number 256 in Lecture Notes in Physics, p. 21, ISBN 978-3-540-16772-3 978-3-540-39847-9

[4] J.F. Berger, M. Girod, D. Gogny, Comput. Phys. Commun. **63**, 365 (1991)

[5] H. Goutte, P. Casoli, J.F. Berger, Nucl. Phys. A **734**, 217 (2004)

[6] H. Goutte, J.F. Berger, P. Casoli, D. Gogny, Phys. Rev. C **71**, 024316 (2005)

[7] D. Regnier, M. Verrière, N. Dubray, N. Schunck, Comput. Phys. Commun. **200**, 350 (2016)

[8] D. Regnier, N. Dubray, N. Schunck, M. Verrière, Phys. Rev. C **93**, 054611 (2016)

[9] D. Regnier, N. Dubray, M. Verrière, N. Schunck, Comput. Phys. Commun. **225**, 180 (2018)

[10] D. Regnier, N. Dubray, N. Schunck, Phys. Rev. C **99**, 024611 (2019)

[11] J. Bartel, P. Quentin, M. Brack, C. Guet, H.B. H{\aa}kansson, Nucl. Phys. A **386**, 79 (1982)

[12] W. Satuła, J. Dobaczewski, W. Nazarewicz, Phys. Rev. Lett. **81**, 3599 (1998)

[13] N. Schunck, J. Dobaczewski, W. Satuła, P. Bączyk, J. Dudek, Y. Gao, M. Konieczka, K. Sato, Y. Shi, X.B. Wang et al., Comput. Phys. Commun. **216**, 145 (2017)

[14] N. Schunck, D. Duke, H. Carr, A. Knoll, Phys. Rev. C **90**, 054305 (2014)

[15] W. Younes, D. Gogny, Phys. Rev. Lett. **107**, 132501 (2011)

[16] D. Regnier, D. Lacroix, G. Scamps, Y. Hashimoto, Phys. Rev. C **97**, 034627 (2018)

[17] W. Younes, D. Gogny, Tech. Rep. LLNL-TR-586678, Lawrence Livermore National Laboratory (LLNL), Livermore, CA (2012)

[18] P. Möller, D. Madland, A.J. Sierk, A. Iwamoto, Nature **409**, 785 (2001)

[19] J. Dobaczewski, W. Nazarewicz, P.G. Reinhard, J. Phys. G: Nucl. Part. Phys. **41**, 074001 (2014)

[20] N. Schunck, J.D. McDonnell, J. Sarich, S.M. Wild, D. Higdon, J. Phys. G: Nucl. Part. Phys. **42**, 034024 (2015)

[21] N. Schunck, J.D. McDonnell, D. Higdon, J. Sarich, S.M. Wild, Eur. Phys. J. A **51**, 1 (2015)

[22] J. Sadhukhan, J. Dobaczewski, W. Nazarewicz, J.A. Sheikh, A. Baran, Phys. Rev. C **90**, 061304 (2014)

[23] J. Zhao, B.N. Lu, T. Nikšić, D. Vretenar, S.G. Zhou, Phys. Rev. C **93**, 044315 (2016)

[24] R. Bernard, S.A. Giuliani, L.M. Robledo, Phys. Rev. C **99**, 064301 (2019)

[25] A. Bulgac, S. Jin, K.J. Roche, N. Schunck, I. Stetcu, Phys. Rev. C **100**, 034615 (2019)

[26] R.N. Perez, N. Schunck, R.D. Lasseri, C. Zhang, J. Sarich, Comput. Phys. Commun. **220**, 363 (2017)

[27] N. Dubray, D. Regnier, Comput. Phys. Commun. **183**, 2035 (2012)

[28] K.T.R. Davies, R.A. Managan, J.R. Nix, A.J. Sierk, Phys. Rev. C **16**, 1890 (1977)

[29] N. Dubray, H. Goutte, J.P. Delaroche, Phys. Rev. C **77**, 014310 (2008)

[30] R.E. Peierls, J. Yoccoz, Proc. Phys. Soc. A **70**, 381 (1957)

[31] F. Villars, Nucl. Phys. A **285**, 269 (1977)

[32] M. Baranger, M. Veneroni, Ann. Phys. **114**, 123 (1978)
[33] P.G. Reinhard, K. Goeke, Rep. Prog. Phys. **50**, 1 (1987)
[34] M. Matsuo, T. Nakatsukasa, K. Matsuyanagi, Prog. Theor. Phys. **103**, 959 (2000)
[35] J. Sadhukhan, K. Mazurek, A. Baran, J. Dobaczewski, W. Nazarewicz, J. Sheikh, Phys. Rev. C **88**, 064314 (2013)
[36] C. Simenel, A.S. Umar, Phys. Rev. C **89**, 031601(R) (2014)
[37] G. Scamps, K. Hagino, Phys. Rev. C **91**, 044606 (2015)
[38] Y. Tanimura, D. Lacroix, G. Scamps, Phys. Rev. C **92**, 034601 (2015)
[39] A. Bulgac, P. Magierski, K.J. Roche, I. Stetcu, Phys. Rev. Lett. **116**, 122504 (2016)
[40] Y. Tanimura, D. Lacroix, S. Ayik, Phys. Rev. Lett. **118**, 152501 (2017)
[41] M. Verrière, N. Schunck, T. Kawano, Phys. Rev. C **100**, 024612 (2019)
[42] R. Bernard, H. Goutte, D. Gogny, W. Younes, Phys. Rev. C **84**, 044308 (2011)
[43] N. Schunck, J.D. McDonnell, D. Higdon, J. Sarich, S. Wild, Nucl. Data Sheets **123**, 115 (2015)
[44] K. Dietrich, J.J. Niez, J.F. Berger, Nucl. Phys. A **832**, 249 (2010)

Employing FREYA for fission product yield evaluations

R. Vogt[1,2,*], *J.* Randrup[3,**], *N.* Vassh[4], *T.* Sprouse[4], and *R.* Surman[4]

[1]Nuclear and Chemical Sciences Division, Lawrence Livermore National Laboratory, Livermore, CA 94551, USA
[2]Physics Department, University of California at Davis, Davis, CA 95616, USA
[3]Nuclear Science Division, Lawrence Berkeley National Laboratory, Berkeley, CA 94720, USA
[4]Department of Physics, University of Notre Dame, Notre Dame, IN 46556, USA

Abstract. The fast event-by-event fission code FREYA (Fission Reaction Event Yield Algorithm) generates large samples of complete fission events while employing only a few physics-based parameters. Not only is FREYA fast, it is also flexible, able to employ a variety of input formats to test the implications of various fission yield evaluations on neutron and photon observables. We describe how FREYA was applied to the neutron-rich nuclei needed for *r*-process nucleosynthesis calculations as an example of this flexibility. Finally, we discuss how we plan to make use of this flexibility to extend FREYA to calculations of cumulative fission product yields to aid evaluations of these yields in the future.

1 Introduction to FREYA

As input, the standard version of FREYA requires the mass distribution of the initial fission fragments, $Y(A)$, and the mean total kinetic energy for a given mass split, $\overline{\mathrm{TKE}}(A)$. These distributions taken from either experimental data or model calculations. There are a number of approaches that can be employed for such calculations, as described elsewhere in these proceedings [1–3]. They are generally successful at reproducing the mass yields, $Y(A)$, but accurate calculations of $\overline{\mathrm{TKE}}(A)$ remain elusive. As discussed in these proceedings [1], density functional theory is being utilized to obtain the fragment excitation energy which could then be used in place of $\overline{\mathrm{TKE}}(A)$ to determine the early state of the fission fragments. These calculations, when more mature, might provide a reliable alternative means to obtain $\overline{\mathrm{TKE}}(A)$ when no data are available, as is often the case, especially for higher excitation energies in general and minor actinides in particular.

In standard FREYA [4], the initial nucleus of mass number A_0 divides into light and heavy fragments, with mass numbers A_L and A_H respectively. A tentative value of the total fragment kinetic energy, TKE, is given by $\overline{\mathrm{TKE}}(A_H)$ and the mean total excitation energy available for rotational and statistical excitation at scission is then $\overline{E}^*_{\rm sc} = Q - \overline{\mathrm{TKE}}$ where $Q = M_0c^2 - M_Lc^2 - M_Hc^2$ is the Q-value for that particular split. The corresponding 'scission temperature' $T_{\rm sc}$ is obtained from $\overline{E}^*_{\rm sc} = a(A_0)T^2_{\rm sc}$ where the scale of the level density parameter $a(A) = A/e_0$, with $e_0 \approx 10\,\mathrm{MeV}$, is an adjustable parameter in FREYA.

The two nascent fission fragments are assumed to be formed with angular momenta perpendicular to the fission axis. In addition to their share of a possible overall rigid rotation

*e-mail: Vogt2@LLNL.gov
**e-mail: JRandrup@LBL.gov

of the fissioning complex, the two angular momenta also acquire thermal fluctuations from the dinuclear wriggling and bending modes. The magnitude of these 'spin fluctuations' is governed by the 'spin temperature' $T_S = c_S T_{\rm sc}$ which can be adjusted through the FREYA parameter c_S. (Thus the spin fluctuations vanish for $c_S = 0$.)

After subtracting the resulting rotational energy of the two fragments, $E_{\rm rot}$, from $\overline{E}^*_{\rm sc}$ the remainder is the combined mean statistical excitation of the two fragments, $\overline{E}_{\rm stat} = \overline{E}^*_{\rm sc} - E_{\rm rot}$. A preliminary partition, $\overline{E}_{\rm stat} = \acute{E}^*_L + \acute{E}^*_H$, is made according to the heat capacities of the fragments, assumed to be proportional to the level density parameters, *i.e.* $\acute{E}^*_L : \acute{E}^*_H = a_L : a_H$. Because the observed neutron multiplicities for actinide nuclei at low energies suggest that the light fragments tend to be disproportionately excited, the light fragment is given a larger excitation energy by the FREYA parameter x, $\overline{E}_L = x\acute{E}_L$, $\overline{E}_H = E_{\rm stat} - \overline{E}_L$, where $x > 1$.

After the mean fragment excitation energies $\overline{E}_i$ have been assigned, FREYA considers thermal fluctuations in the statistical excitation. The mean fragment excitation is related to the fragment temperature T_i by $\overline{E}_i = a_i T_i^2$ with the associated variance being $\sigma^2_{E_i} = 2\overline{E}_i T_i$. An energy fluctuation δE^*_i is sampled from a truncated normal distribution of variance $2c\overline{E}_i T_i$ and the excitations are adjusted accordingly, $E^*_i = \overline{E}_i + \delta E^*_i$, $i = L, H$. Energy is conserved by a compensating opposing fluctuation in TKE, $\mathrm{TKE} = \overline{\mathrm{TKE}} - \delta E^*_L - \delta E^*_H$. The factor c multiplying the variance is another adjustable parameter in FREYA with c typically larger than unity. The parameter compensates for the truncation of the normal distribution due to energy conservation. We note that for spontaneous fission of some isotopes with low average neutron multiplicities but relatively broad multiplicity distributions $P(\nu)$, c can be as large as 3 [5]. Finally, to reproduce the average neutron multiplicity, $\overline{\nu}$, the overall average TKE may be adjusted by a fifth FREYA parameter, dTKE.

The values of these five explicit parameters in the standard version of FREYA, e_0, c_S, x, c and dTKE, can be adjusted to data. Three are tightly correlated with specific observables, having little to no direct effect on others: The photon emission depends directly on c_S; the parameter x, governing statistical excitation energy sharing, determines both the neutron multiplicity as a function of fragment mass, $\nu(A)$, and the angular correlations between neutrons emitted during a fission event; and c influences the neutron multiplicity distribution $P(\nu)$ and its moments. The only observable directly affected by e_0 is the prompt fission neutron spectrum (PFNS), though the other parameters also exhibit some influence over the PFNS. The parameter dTKE, while related to the other parameters, serves primarily to ensure that $\overline{\nu}$ matches the evaluated value. Details about the parameter values and their correlations are given in Ref. [5].

After the fragments have been characterized, neutron and photon emission follows. The neutrons are evaporated isotropically in the frame of the emitting fragment, apart from a slight flattening due to the fragment rotation [6]. Their energy is sampled from a black-body energy spectrum, $dN_n/dE_n \sim E_n \exp(-E_n/T_{\rm max})$, where $T_{\rm max}$ is the maximum possible temperature in the daughter nucleus. FREYA generally assumes that neutron evaporation continues until the nuclear excitation energy is below the neutron separation energy $S_{\rm n}$, so that neutron evaporation continues as long as energetically possible.

After neutron evaporation has ceased, the excited product nucleus emits photons. First, statistical photons are emitted isotropically with an energy distribution sampled from a black-body spectrum modulated by a giant-dipole resonance form factor. When the nuclear excitation energy enters the regime of the tabulated decays from the RIPL-3 compilation [7], FREYA switches to a discrete cascade which is continued until the half-life exceeds a specified value, $t_{\rm max}$, based on the detector response time, or until the nucleus is in its ground state. The average photon energy and photon multiplicity depend on $t_{\rm max}$ as well as on the minimum photon energy measurable in the detectors, denoted in FREYA as $g_{\rm min}$. While the photon observables depend on both $t_{\max}$ and $g_{\min}$, these quantities are not parameters but depend on specific experimental details.

It is clear that while FREYA has the five physics-based adjustable parameters described above, there are also many implicit inputs that rely on models and data, including fission barriers, nuclear masses, photon strength functions and the input yields themselves. For a study of the FREYA sensitivity to the input yield functions, see Ref. [8]. Because of its speed and adjustable formats for inputs, one of which is described in Sec. 2, FREYA can be used to test input yields and excitation energies calculated with models. Including data for delayed emission, FREYA can also be used to simulate cumulative yields in different time windows as well as provide a way to track the delayed emission in addition to prompt emission, as is outlined in Sec. 3.

2 FREYA in r-process nucleosynthesis calculations

While FREYA has been released for only a dozen isotopes, many more are required for calculating nuclear abundances in the r process. See Ref. [9] for a more thorough discussion of the astrophysics processes. The discussion here will focus only on the role of FREYA in such calculations. As Ref. [9] made clear, fission fragments can be important contributors to the nucleosynthesis abundance patterns for neutron-rich astrophysical sites such as neutron star mergers. The neutrons emitted by the fission fragments may also be important, especially at later times, because they contribute to neutron capture,

In the r process, fission can be initiated several ways: neutron-induced, beta-delayed, and spontaneous. Different fission processes are important at different time periods in the r process. The different types of fission also occur at different compound nucleus excitation energies. To take these fission initiators and excitation energies into account, it is necessary to have more accurate calculations of the fission fragment mass distributions and the neutrons emitted during their de-excitation for models of astrophysical processes. Models are required because data for fission of high-Z neuron-rich nuclei involved in fission recycling do not exist and are likely to remain unavailable. Heretofore, more naive treatments have been employed, including 50/50 splits and semi-empirical formulas. Previously, treating the yields and neutron emission from spontaneous, neutron-induced, and beta-delayed fission as distinct by taking into account their unique excitation energy dependence had not been done.

The work in Ref. [9] changed that. The 2016 version of the GEF code [10], which uses systematics in Z, A and excitation energy of the compound nucleus based on data and models to calculate the fission fragment yields and excitation energies for neutron emission from the fragments, was employed. FREYA was also employed in the study to quantify the uncertainties in the resulting abundance pattern based on the de-excitation processes in the two codes. To do this, FREYA had to be modified to accept the GEF yields: $Y(A, Z)$, $Y(A)$ and $Y(\text{TKE})$.

Before proceeding, we investigated two test cases, ^{252}Cm(sf) and ^{258}Pu(sf), to see if FREYA could successfully incorporate the GEF yields and produce reasonable results for neutron-rich nuclei, far outside the regime where it had been previously applied. Results for the test cases are shown in Fig. 1.

The results in Fig. 1 showed that FREYA could take the GEF yields and faithfully reproduce them. It also shows that there are some differences in the fission product yields obtained with the two codes, due to the differences in the sharing of the excitation energy between the fragments. This difference is easier to see in the neutron multiplicity as a function of fragment mass. While $\bar{\nu}(A)$ from the two codes is in rather good agreement for the default parameter of x used in the FREYA calculation of ^{252}Cm(sf) in the range $100 < A < 150$, there are discrepancies for the more asymmetric splits. This is not always the case, as seen

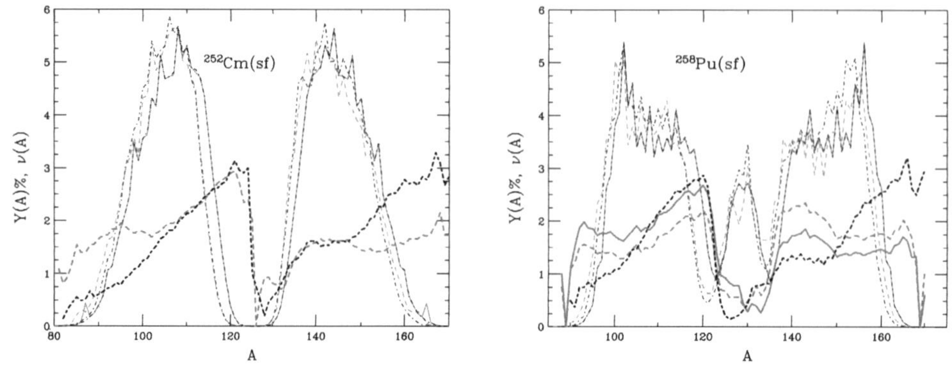

Figure 1. The fission fragment and fission product mass yields as well as the mean neutron multiplicity as functions of fission fragment mass number A for ^{252}Cm(sf) (left) and ^{258}Pu(sf) (right). The thin solid red curves are the fission fragment yields from FREYA based on the GEF yields, underneath the red curves, in black. The thin dashed red curves are the FREYA fission product yields, after prompt neutron emission, while the thin black dot-dashed curve is the GEF result. The $Y(A)$ results are given in percent and are normalized to two fragments per fission, or 200 %. The thick curves show the mean neutron multiplicity as a function of the primary fission fragment mass. The red curves show the FREYA results for $\bar{\nu}(A)$, while the black curve shows the GEF results. Note that the calculations for ^{258}Pu(sf) (right-hand panel) show two results for FREYA. The dashed curve uses a default value for the excitation energy sharing parameter x, $x = 1.15$, also used for the Cm calculation on the left, while the solid curve uses $x = 1.4$, fixed to match the GEF value of the average neutron multiplicity from GEF. Note that the FREYA $Y(A)$ for fission product yields is obtained using $x = 1.4$.

for ^{258}Pu(sf) where the value of x used in FREYA had to be increased to 1.4 to reproduce the average neutron multiplicity $\bar{\nu}$ from GEF.

Because tuning parameters to reproduce the GEF results would provide little new information while giving no hint of uncertainties in the results from the two codes, we used the parameters for ^{252}Cf for spontaneous fission and ^{239}Pu for neutron-induced and beta-delayed fission. We found that FREYA generally resulted in higher neutron emission from fission than GEF, sometimes 2-3 more neutrons per fission. The difference is largest in the most neutron-rich nuclei on the nuclear chart and comes mostly from the light fragment. This could be due to the fact that neutron emission from GEF is internally consistent, albeit with large uncertainties, while although FREYA starts from the same yields it handles the excitation energy sharing differently than does GEF. These differences can result in 10-20% relative changes in the abundance pattern between FREYA and GEF. See Ref. [9] for full details.

3 FREYA for Cumulative Yields

The current version of FREYA simulates spontaneous or induced fission from the pre-compound stage through prompt neutron and photon emission to the resulting product nuclei. These are generally beta unstable and, consequently, on a longer time scale, they will undergo a sequence of beta decays accompanied by photon radiation. We plan to incorporate these delayed processes into FREYA to be able to simulate the decays leading to cumulative fission yields. With this information incorporated, we can compare to fission product yield measurements at different time scales, from seconds to minutes to days. Because FREYA simulates complete events, it is also possible to follow the trajectories of the delayed particles that are emitted, including the neutrons, photons, electrons and antineutrinos. (Note that for the *r*-

process application described in Sec. 2, delayed emission from fission products was treated in the nucleosynthesis network calculation using the decay data described in Ref. [9].

If the needed information about the delayed processes is available, then the effort required to incorporate these processes into the existing version of FREYA should be relatively manageable through a new subroutine called after the prompt processes have ceased. This routine will follow the delayed decay processes in a Monte-Carlo manner and, for each product nucleus. it will provide both the resulting beta-stable fission fragment daughter and the kinematic information about the emitted photons, and leptons (electrons and antineutrinos). It should be relatively straightforward to construct and implement such a subroutine without slowing FREYA down by any appreciable degree.

The FIER code package [11] contains essentially all the known information on the beta-delayed decay processes. In FIER, the available information is tabulated in two files: one containing the half-lives and associated branching ratios and the other the beta-decay end point energies and associated intensities. The N, Z domain covered by these data files are illustrated in black and red respectively on the left-hand side of Fig. 2. However, some portion of the beta-decay data is not available and therefore the incorporation of these processes must rely on some modeling. We do note that many of the fission products from spontaneous fission of ^{252}Cf do have data available so that modeling is not required for every possible fission product. This does, of course, depend on the fissioning isotope as well as its excitation energy.

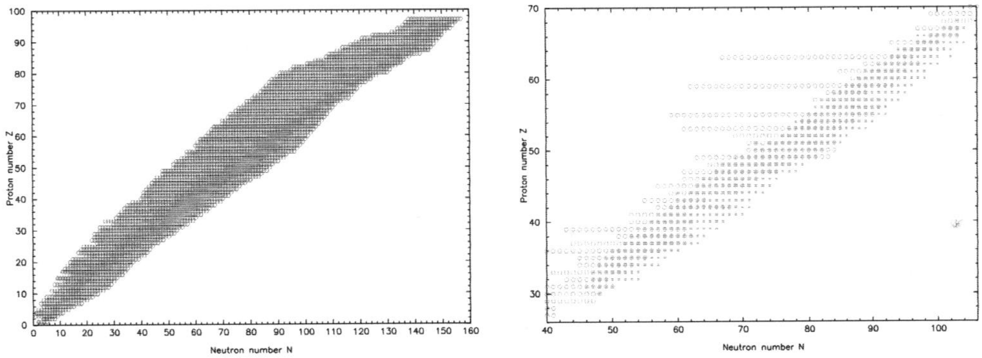

Figure 2. (Left) The (N, Z) domains included in the FIER data tables. The dark bins show the nuclides with branching ratios and half-lives included, while the overlaid red bins show those with information on beta-decays. Clearly there is beta-decay information available for far fewer nuclides. The (N, Z) distribution of the fission products from spontaneous fission of ^{252}Cf(sf) is shown by the green bins. It is apparent that most but not all of the nuclei of interest are contained in the FIER tables. (Right) Magnified views of the relevant parts of the (N, Z) plane for ^{252}Cf(sf). The half-life information is omitted because all fission fragments of interest are included. Note that the green symbols show the location of fission products from one million spontaneous-fission events.

Because accurate modeling of the beta-delayed decay processes is difficult, it is unrealistic to imagine that a fully satisfactory theoretical model of beta-delayed decay chains will be available in the near future. However, the state of the art is steadily improving.

It is therefore preferable to start with a relatively simple form of modeling while ensuring that the augmented version of FREYA is structured so that model replacements can readily be made. It will then be relatively easy to maintain a state-of-the-art treatment while also allowing comparisons between alternative decay models.

4 Summary

FREYA is a robust, fast, flexible tool for studying complete fission events. While it currently handles only prompt emission, allowing access only to the independent fission product yields, it can be extended to include delayed emission by incorporating the FIER data, giving access to the cumulative fission product yields as well.

Acknowledgments

We thank R. Surman and N. Vassh for collaboration on the *r*-process calculations. The work of R.V. was performed under the auspices of the U.S. Department of Energy by Lawrence Livermore National Laboratory under Contract DE-AC52-07NA27344. The work of J.R. was performed under the auspices of the U.S. Department of Energy by Lawrence Berkeley National Laboratory under Contract DE-AC02-05CH11231. This work was supported by the Office of Defense Nuclear Nonproliferation Research & Development (DNN R&D), National Nuclear Security Administration, U.S. Department of Energy.

References

[1] N. Schunck, these proceedings.
[2] M. Verriere, these proceedings.
[3] I. Abdurrahman, these proceedings.
[4] J. M. Verbeke, J. Randrup and R. Vogt, Comp. Phys. Comm. **222**, 263 (2018).
[5] J. T. Van Dyke, L. A. Bernstein and R. Vogt, Nucl. Instrum. Meth. A **922**, 36 (2019).
[6] J. Randrup and R. Vogt, Phys. Rev. C **89**, 044601 (2014).
[7] R. Capote *et al.*, Nucl. Data Sheets **110**, 3107 (2009).
[8] J. Randrup, P. Talou and R. Vogt, Phys. Rev. C **99**, 054619 (2019).
[9] N. Vassh *et al.*, J. Phys. G **46**, 065202 (2019).
[10] K.-H. Schmidt, B. Jurado, C. Amouroux and C. Schmitt, Nucl. Data Sheets **131**, 107 (2016).
[11] E. Matthews *et al.*, Nucl. Istrum. Meth. A **891**, 111 (2018).

Using CGMF to estimate corrections for fission yields measured via γ-ray spectroscopy

P. Jaffke[1,2,*], *P.* Talou[3], *M.* Devlin[4], and *N.* Fotiades[4]

[1]System Evaluation Division, Institute for Defense Analyses, Alexandria, VA 22311, USA
[2]Theoretical Division, Los Alamos National Laboratory, Los Alamos, NM 87545, USA
[3]XCP-5, Computational Physics Division, Los Alamos National Laboratory, Los Alamos, NM 87545, USA
[4]Physics Division, Los Alamos National Laboratory, Los Alamos, NM 87545, USA

Abstract. Fission product yields have been inferred using γ-ray spectroscopy for several decades. Typically, these efforts have focused on even-Z even-A fission products as their nuclear structure are less complicated. To further simplify the situation, it is often assumed that no side-feeding to the ground-state occurs and multiplicity cuts have a negligible effect on the inferred yields. Using CGMF, a Hauser-Feshbach statistical decay model for the primary fission fragments, we estimate the impact of these assumptions and determine corrections for specific fission product yields. We report on these corrections and investigate their sensitivity to various nuclear parameters, specifically the spin distribution of the fission fragments and the assumed nuclear structure. Our results indicate that even in the simplest of cases, say the $2^+ \rightarrow 0^+$ transitions in even-Z even-A fragments, average level corrections are on the order of 75%.

1 Introduction

Gamma-ray spectroscopy has been utilized to infer fission yields for a variety of nuclear reactions [1–4]. Experimental efforts have spanned prompt γ-ray emission times from within $1\,\mu$s or so [5, 6] to emissions after β-decays in the days to months range [7, 8]. Prompt γ-ray emissions are sensitive to the independent fission product yields (i.e. prior to β-decays), while later γ-ray emissions after β-decay are more sensitive to cumulative yields. Here we focus on γ-ray emissions prior to 150 ns.

Often, experimental assumptions and corrections must be applied to the data in order to properly account for energy resolution, nuclear transition strengths, and multiplicity cut effects. In particular, for prompt γ rays, fission products are generated with a distribution in excitation energy and spin. This distribution further complicates the necessary corrections. In this work, we present preliminary calculations of three potential corrections required for prompt γ-ray spectroscopy and outline a procedure for a given experimental setup.

The three considered corrections are dubbed the purity ϵ_P, level ϵ_L, and multiplicity ϵ_M corrections. The purity correction stems from γ-ray transitions of similar energy to the desired transition being included in the finite energy bin of a given γ-ray spectroscopy detector.

*e-mail: pjaffke@ida.org

The level correction originates from the branching ratios of the different emissions. The multiplicity correction arises when experimental cuts are applied to either reduce the data rate or provide more assurance of a fission coincidence.

In Sec. 2, we describe the Hauser-Feshbach de-excitation model, which is used to estimate the needed corrections. Next, in Sec. 3, we detail the three specific corrections related to prompt γ-ray spectroscopy. In Sec. 4, the results are presented to illustrate the impact of using the corrections. Lastly, we conclude in Sec. 5.

2 Hauser-Feshbach de-excitation

Fission fragments are de-excited with a Monte Carlo statistical decay following the theory of Hauser and Feshbach [9]. This implementation is given by the `CGMF` [10] code, described in detail in Ref. [11]. The de-excitation begins with a provided distribution of fission fragments

$$Y(A, Z, \mathrm{TKE}, J^{\pi}) = Y(A) \times Y(\mathrm{TKE}|A) \times Y(Z|A) \times P(J^{\pi}|A, Z, U). \tag{1}$$

In Eq. 1, the fission fragment yield in mass A, charge Z, total kinetic energy TKE, and spin-parity J^{π} is given by $Y(A, Z, \mathrm{TKE}, J^{\pi})$. In the calculations presented here, we utilize fits to experimental data for the $^{235}\mathrm{U}(n_{\mathrm{th}}, f)$ reaction. A global fit is applied to $Y(A)$ and $Y(\mathrm{TKE}|A)$ experimental data, as done in Ref. [12], while the $Y(Z|A)$ is given by the Wahl systematics [13].

First, yields are sampled for a given fission event from Eq. 1 and the total excitation energy TXE is calculated via energy conservation $Q = \mathrm{TKE} + \mathrm{TXE}$. The Q-value is determined from nuclear masses [14], the incident neutron energy, and the binding energy. Next, the TXE is shared between the two fragments with a ratio of temperatures R_T, given by

$$R_T^2 = \frac{T_l^2}{T_h^2} \approx \frac{U_l a_h}{U_h a_l}, \tag{2}$$

where T_i is the nuclear temperature, U_i is the excitation energy, and a_i is the level density parameter for the light or heavy fragment. The use of Eq. 2 implicitly assumes the fragments are produced with sufficient excitation energy such that the level density is well-represented by a Fermi-gas model. In `CGMF`, we use a mass-dependent $R_T \equiv R_T(A)$, where the value of $R_T(A)$ is determined from fits to experimental $\bar{\nu}(A)$ data [15].

Next, the spin-parity distribution is sampled on an event-by-event basis. In `CGMF`, a spin-cutoff distribution $P(J^{\pi})$ is used

$$P(J^{\pi}) = N(2J + 1) \exp\left[\frac{-J(J + 1)\hbar^2}{2\alpha T \mathcal{I}_0(A, Z)}\right], \tag{3}$$

where $\mathcal{I}_0(A, Z)$ is the moment of inertia for a rigid rotor, using the ground-state deformations of Ref. [14]. The factor N is a normalization factor and α is a spin scaling factor used to control the competition between γ-ray and neutron emission. For the calculations presented here, we use $\alpha = 1.45$ for $^{235}\mathrm{U}(n_{\mathrm{th}}, f)$. In `CGMF`, we assume an equal probability of parities and assume even(odd)-A fragments have even(odd)-J spin. Figure 1 illustrates the distribution of the primary fragments (i.e. prior to neutron emission) in blue for $^{100}\mathrm{Zr}$ from the $^{235}\mathrm{U}(n_{\mathrm{th}}, f)$ reaction. The curve denotes using Eq. 3 and provides a good representation of the sampled distribution (points). The orange is the combined distribution from all events resulting in $^{100}\mathrm{Zr}$ as a fission product (i.e. after prompt neutron emission). The neutron emission itself does not remove significant angular momentum, so a linear combination of Eq. 3 can be constructed to provide a reasonable estimation of the sampled points.

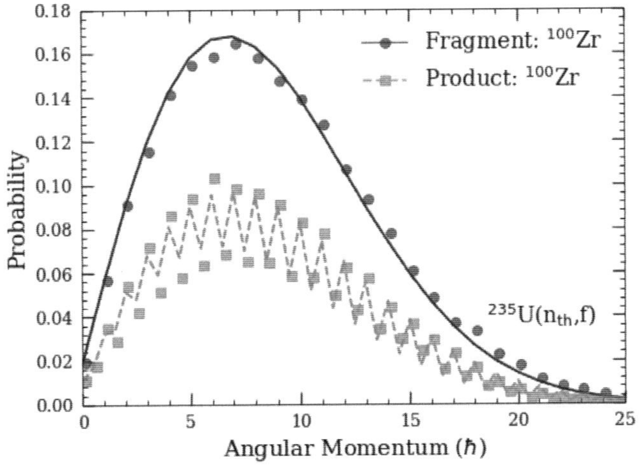

Figure 1. (Color online) Spin distribution of CGMF events initially sampling a ^{100}Zr fragment (in blue) and events resulting in a ^{100}Zr product after prompt neutron emission (in orange). Points denote the calculated distributions by CGMF, while the lines utilize Eq. 3. For the resulting product, a linear combination of $P(J)$ weighted by the prompt neutron emission is used to determine the total $P(J)$. Results are for ^{235}U(n_{th}, f) and one million events.

Now that an initial distribution $Y(A, Z, U, J^\pi)$ has been established using Eqs. 1–3, the de-excitation can be determined through the calculation of transmission coefficients for γ-ray and neutron emission. As this process has been well-documented before [10, 11], we only summarize here. In CGMF, neutron transmission coefficients are determined using an optical model calculation with a global optical potential from Ref. [16]. Gamma-ray transmission coefficients follow the strength-function formalism of Ref. [17], with parameters from the Reference Input Parameter Library (RIPL-3) [18]. Discrete levels and their branching ratios are also from RIPL-3. The level densities follow the Gilbert-Cameron [19] form, with energy-dependent level density parameters via Ref. [20].

3 Corrections for prompt γ rays

Several calculations were conducted to estimate the effect of nuclear data and the spin scaling factor α on the determined corrections. The first calculation uses the 'optimal' spin scaling factor α discussed above and the improved 2015 RIPL-3 database. The second calculation supplements RIPL-3 with discrete level data for ^{104}Zr [21] and ^{138}Te [22, 23]. The third calculation increases the spin scaling factor α by about 10%. All calculations use one million events. Next, we detail the three corrections considered for prompt γ-ray spectroscopy.

The purity correction ϵ_P accounts for γ rays, both discrete and continuum transitions, mistakenly being counted in the photopeak of the target γ-ray energy E_γ. This quantity is determined by collecting all CGMF fission events containing γ rays with energy $E_\gamma \pm \delta(E_\gamma)/2$, where $\delta(E_\gamma)$ is a typical energy resolution of a Ge detector [24]. The fraction of those events that are actually emitted from the target fission product is given by ϵ_P. An example for several even-Z even-A fission products of ^{235}U(n_{th}, f) are given in the left-hand plot of Fig. 2.

The level correction ϵ_L accounts for the branching ratios and, implicitly, the distribution in excitation energy and spin-parity. For this quantity, we mask on all CGMF events that generated the target fission product. Next, we determine the fraction of this subset that actually

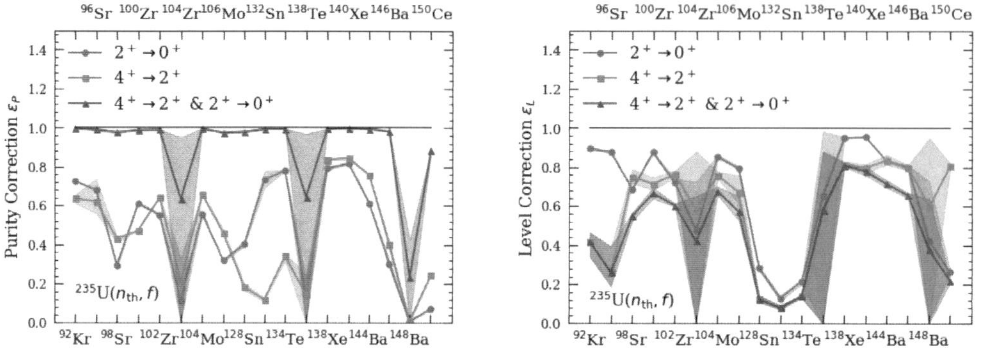

Figure 2. (Color online) Left plot shows the purity correction ϵ_P, while the right plot shows the level correction ϵ_L ratio for several even-Z even-A fission products. Circles (green) indicate the $2^+ \to 0^+$ transition, squares (blue) indicate the $4^+ \to 2^+$ transition, and triangles (red) indicate a double-gate on both. The curves and points signify the mean correction over the three CGMF calculations (see text). The bands show the upper and lower correction values, where bands that have a lower bound of zero indicate that the particular γ-ray transition did not exist in the default RIPL-3 database.

produced the target γ-ray transition E_γ. It is commonly assumed that for even-Z even-A fission products the $2^+ \to 0^+$ transition *always* occurs. In the right-hand plot of Fig. 2, we show that this is typically not the case. Rather, the median value is closer to 75%.

We note two significant findings from Fig. 2 here. First, the corrections depend on both the included nuclear data and the spin scaling factor α. Second, the corrections depend on the specific fission product transition greatly. Another significant finding, not displayed in Fig. 2, is that the corrections depend on the fission reaction, especially ϵ_P which is sensitive to the yields. Thus, it is important to realize that the exact corrections presented here are specific

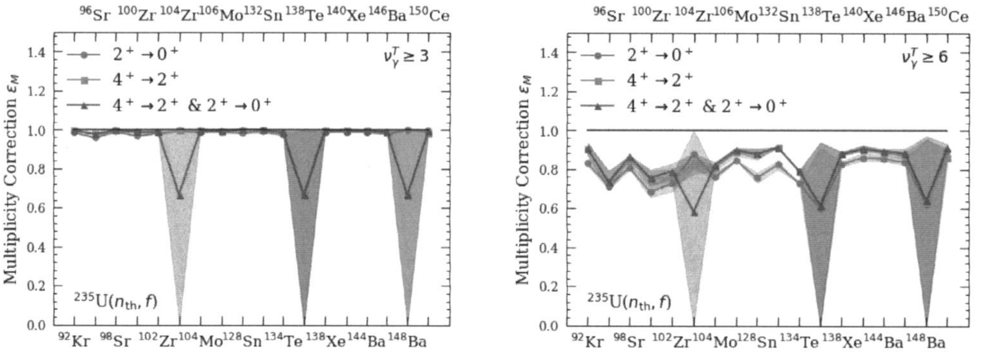

Figure 3. (Color online) Multiplicity correction ratio ϵ_M for $\nu_\gamma^T \geq 3$ (left) and $\nu_\gamma^T \geq 6$ (right) for several even-Z even-A fission products. Circles (green) indicate the $2^+ \to 0^+$ transition, squares (blue) indicate the $4^+ \to 2^+$ transition, and triangles (red) indicate a double-gate on both. The curves and points signify the mean correction over the three CGMF calculations (see text). The bands show the upper and lower correction values, where bands that have a lower bound of zero indicate that the particular γ-ray transition did not exist in the default RIPL-3 database.

to the assumed specifications (detector resolution, fission reaction, etc.), so a similar exercise should be conducted on an experiment-by-experiment basis.

The third correction is the multiplicity correction ϵ_M shown in Fig. 3, which accounts for the bias introduced by limiting the total number of γ rays used to identify a fission event. This is often used to improve the signal-to-noise ratio or ensure that γ rays are coincident with a fission event. In our analyses, we assume no ability to distinguish which fission product emitted the γ ray, a different approach than used previously [25]. Thus, our multiplicity cuts are for the total fission event. In Fig. 3, we note that a cut of $\nu_\gamma^T \geq 3$ has little effect, whereas a cut of $\nu_\gamma^T \geq 6$ requires a correction of about 80%, but there is considerable variance between the fission products and transitions. We note that we have not considered the detector efficiency here.

We determine ϵ_M by first masking on all `CGMF` events producing the target fission product. We apply a second mask to these sub-events that generate all target γ-ray transitions, so as not to double-count the level correction. We then determine the total number of γ rays emitted in these double-masked events and compare against potential cuts of $\nu_\gamma^T = 3, 6, 9$ γ rays. A notable result from this calculation, seen in Fig. 3, is that ϵ_M depends strongly on the fission product (i.e. a γ-ray multiplicity cut disproportionately affects certain fission products).

4 Results

Figure 4 illustrates the effect of utilizing these corrections. The (red) open squares show the inferred yields from a summation of the $4^+ \rightarrow 2^+$ transition photopeak over all `CGMF` events for each even-Z even-A fission product for $^{235}U(n_{th}, f)$. These results take into account the typical energy resolution of a Ge detector and utilize a $\nu_\gamma^T \geq 3$ multiplicity cut, a conservative estimate in Ge detector arrays not spanning a 4π solid angle [26]. The solid curve denotes the

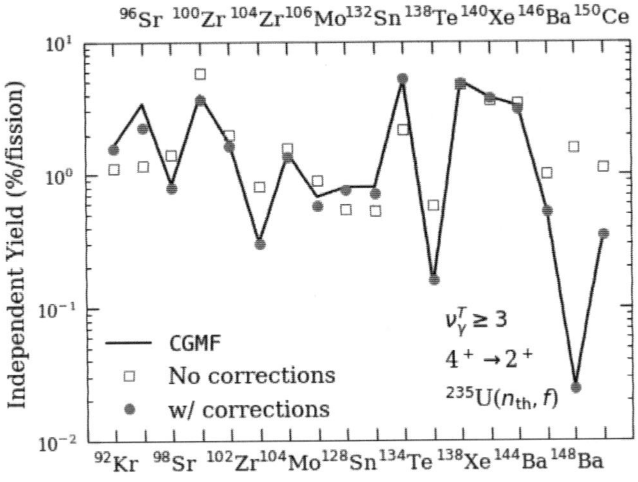

Figure 4. (Color online) Inferred yields for several even-Z even-A fission products from $^{235}U(n_{th}, f)$ using a single-gate on the $4^+ \rightarrow 2^+$ transition. Open (red) squares denote a simple photopeak area sum accounting for a typical [24] Ge detector resolution and a total γ-ray multiplicity cut of $\nu_\gamma^T \geq 3$. Closed (green) circles apply the average corrections over the three `CGMF` calculations (see text) to the photopeak sum results. The line shows the 'true' fission product yields sampled by `CGMF`.

'true' yields that comprise the sampling basis in Eq. 1. After applying the corrections

$$Y_{\text{corr}}(A,Z) = Y_{\text{inf}}(A,Z) \times \left(\frac{\epsilon_P}{\epsilon_L \epsilon_M} \right), \tag{4}$$

to the inferred yield $Y_{\text{inf}}(A,Z)$, the (green) circles denote the corrected yields $Y_{\text{corr}}(A,Z)$. The corrections ϵ_P, ϵ_L, and ϵ_M used in Fig. 4 are the average correction over the three `CGMF` calculations (i.e. the curves and points in Fig. 2 and Fig. 3).

We note that, for the most part, the corrected yields are closer to the 'true' yields by a significant margin and similar results were found for different transitions, fission reactions, and multiplicity cuts. A similar strategy was applied to GEANIE data for the ^{238}U(n=1.75 MeV, f) reaction [25]. The resulting yields were in reasonable agreement with inverse kinematics results [27].

5 Conclusion

We have outlined a procedure for calculating corrections to fission product yields inferred through γ-ray spectroscopy, a powerful but complex procedure. Three corrections were considered. The first results from including γ-ray transitions emitted from other fission products in an energy bin of a given detector. The second stems from the nuclear structure and distributions of excitation energy and spin of the generated fission products. The last considers implicit biases introduced by applying γ-ray multiplicity cuts. We used a Monte Carlo implementation of the statistical Hauser-Feshbach decay theory, the `CGMF` code, to infer the values of these corrections.

The corrections were found to improve the inferred yields from Doppler-corrected photopeak areas alone. We have calculated the value of these corrections and demonstrated their impact and variance with respect to model parameters. We emphasize that the exact values determined here incorporate a specific detector resolution and fission reaction. Thus, it will be important to conduct a similar analysis presented here for each detector setup and experiment.

This work was supported by the Office of Defense Nuclear Nonproliferation Research & Development (DNN R&D), National Nuclear Security Administration, US Department of Energy. It was performed under the auspices of the National Nuclear Security Administration of the US Department of Energy at Los Alamos National Laboratory under Contract DEAC52-06NA25396.

References

[1] I. Ahmad, W.R. Phillips, Reports on Progress in Physics **58**, 1415 (1995)

[2] J. Hamilton, A. Ramayya, S. Zhu, G. Ter-Akopian, Y. Oganessian, J. Cole, J. Rasmussen, M. Stoyer, Progress in Particle and Nuclear Physics **35**, 635 (1995)

[3] N. Fotiades, J.A. Cizewski, K.Y. Ding, R. Krücken, J.A. Becker, L.A. Bernstein, K. Hauschild, D.P. McNabb, W. Younes, P. Fallon et al., Physica Scripta **2000**, 127 (2000)

[4] A. Bogachev, L. Krupa, O. Dorvaux, E. Kozulin, M. Itkis, M.G. Porquet, A. Astier, D. Curien, I. Deloncle, G. Duchene et al., The European Physical Journal A **34**, 23 (2007)

[5] G.M. Ter-Akopian, J.H. Hamilton, Y.T. Oganessian, J. Kormicki, G.S. Popeko, A.V. Daniel, A.V. Ramayya, Q. Lu, K. Butler-Moore, W.C. Ma et al., Phys. Rev. Lett. **73**, 1477 (1994)

[6] S. Mukhopadhyay, L.S. Danu, D.C. Biswas, A. Goswami, P.N. Prashanth, L.A. Kinage, A. Chatterjee, R.K. Choudhury, Phys. Rev. C **85**, 064321 (2012)
[7] M. Gooden, C. Arnold, J. Becker, C. Bhatia, M. Bhike, E. Bond, T. Bredeweg, B. Fallin, M. Fowler, C. Howell et al., Nuclear Data Sheets **131**, 319 (2016), special Issue on Nuclear Reaction Data
[8] Krishichayan, M. Bhike, C.R. Howell, A.P. Tonchev, W. Tornow, Phys. Rev. C **100**, 014608 (2019)
[9] W. Hauser, H. Feshbach, Physical Review **87**, 366 (1952)
[10] P. Talou, T. Kawano, I. Stetcu, LA-CC-13063, Los Alamos Nat. Lab. (2013)
[11] B. Becker, P. Talou, T. Kawano, Y. Danon, I. Stetcu, Phys. Rev. C **87**, 014617 (2013)
[12] P. Jaffke, P. Möller, P. Talou, A.J. Sierk, Phys. Rev. C **97**, 034608 (2018)
[13] A.C. Wahl, Tech. rep., Los Alamos Nat. Lab. LA-13928 (2002)
[14] P. Möller, A. Sierk, T. Ichikawa, H. Sagawa, Atomic Data and Nuclear Data Tables **109-110**, 1 (2016)
[15] Vorobyev, A.S., Shcherbakov, O.A., Gagarski, A.M., Val'ski, G.V., Petrov, G.A., EPJ Web of Conferences **8**, 03004 (2010)
[16] A. Koning, J. Delaroche, Nuclear Physics A **713**, 231 (2003)
[17] J. Kopecky, M. Uhl, Phys. Rev. C **41**, 1941 (1990)
[18] R. Capote, M. Herman, P. Obložinskỳ, P. Young, S. Goriely, T. Belgya, A. Ignatyuk, A. Koning, S. Hilaire, V. Plujko et al., Nuclear Data Sheets **110**, 3107 (2009), special Issue on Nuclear Reaction Data
[19] A. Gilbert, A.G.W. Cameron, Canadian Journal of Physics **43**, 1446 (1965)
[20] A. Ignatyuk, K. Istekov, G. Smirenkin, Sov. J. Nucl. Phys.(Engl. Transl.);(United States) **29** (1979)
[21] E.Y. Yeoh, S.J. Zhu, J.H. Hamilton, A.V. Ramayya, Y.X. Liu, Y. Sun, J.K. Hwang, S.H. Liu, J.G. Wang, Y.X. Luo et al., Phys. Rev. C **82**, 027302 (2010)
[22] P. Lee, C.B. Moon, C.S. Lee, A. Odahara, R. Lozeva, A. Yagi, S. Nishimura, P. Doornenbal, G. Lorusso, P.A. Söderström et al., Phys. Rev. C **92**, 044320 (2015)
[23] W. Urban, K. Sieja, T. Rząca-Urban, M. Czerwiński, H. Naïdja, F. Nowacki, A.G. Smith, I. Ahmad, Phys. Rev. C **93**, 034326 (2016)
[24] A. Owens, Nuclear Instruments and Methods in Physics Research Section A: Accelerators, Spectrometers, Detectors and Associated Equipment **238**, 473 (1985)
[25] N. Fotiades, P. Casoli, P. Jaffke, M. Devlin, R.O. Nelson, T. Granier, P. Talou, T. Ethvignot, Phys. Rev. C **99**, 024606 (2019)
[26] J.N. Wilson, M. Lebois, L. Qi, P. Amador-Celdran, D. Bleuel, J.A. Briz, R. Carroll, W. Catford, H. De Witte, D.T. Doherty et al., Phys. Rev. Lett. **118**, 222501 (2017)
[27] D. Ramos, M. Caamaño, A. Lemasson, M. Rejmund, L. Audouin, H. Álvarez-Pol, J.D. Frankland, B. Fernández-Domínguez, E. Galiana-Baldó, J. Piot et al., Phys. Rev. Lett. **123**, 092503 (2019)

Correlated transitions in symmetric and dominant fission modes, and multipole moments of fission fragments

Satoshi Chiba[1], *Mark D. Usang*[1,2], *Chikako Ishizuka*[1] *Fedir Ivanyuk*[1,3] *and Zhang Xuan*[1]

[1]Tokyo Institute of Technology, 152-8550, Tokyo, Japan
[2]Malaysia Nuclear Agency, 43000, Bangi, Malaysia
[3]Institute of Nuclear Research, 03028, Kiev, Ukraine

Abstract. Systematic and anomalous trends in fragment mass and TKE (total kinetic energy) distributions are investigated in terms of 4D Langevin model developed at Tokyo Tech. We have found that correlated transitions in symmetric components and dominant modes (symmetric v.s. asymmetric) can explain the prominent systematic and anomalous features of fission observables. We have also elucidated that interplay between spherical and deformed magicity at A=132 and A=142 to 144 is important in both observables.

1 Introduction

Fission is the most important and fundamental physics process in nuclear applications, but it is important also in r-process nucleosynthesis in binary neutron-star merger as fission recycling is now believed to be taking place. Recently, it is strongly recognized that signals from fission (or fission products) plays essential role in nuclear security and safeguards. Thanks to recent progress in experimental technique, nuclear theory and computer science, our understanding on nuclear fission is in a stage of rapid progress, although it is still a mysterious phenomenon due to its nature as a large-amplitude collective motion of many nucleon systems.

Recently, we have carried out a systematic effort to construct a computational framework of nuclear fission, starting from compound nuclei through scission to emission prompt neutrons till β-decay, with less number of parameters and/or hypotheses. Now, our main interest is to connect the outcome of 4-D Langevin approach[1] to prompt neutron emission to predict multiplicities of prompt neutrons as a function of mass number of fission fragments, the so-called saw-tooth structure.

In this work we wish to describe our recent work with the 4D Langevin model developed at Tokyo Institute of Technology. Here, it will be explained how the systematic and anomalous trends in the fragment mass and TKE distributions can be explained as a correlated transition of symmetric mode from super-long to super short, and the dominant mode lying at the asymmetric mode, shifting to symmetric, and then asymmetric modes again as $Z^2/A^{1/3}$ of the fissioning nuclei increases. Then, we calculated $\langle Q_{20} \rangle$ and $\langle Q_{30} \rangle$ of fission fragments, and obtained a direct evidence of the super-long mode for the symmetric components in the fission of actinide region, and large $\langle Q_{30} \rangle$ for mass region of 142-144 corresponding to softness to the octupole deformation, which is traditionally interpreted as originated from deformed shell.

2 Method

We describe the fission process within the Langevin approach [2], i.e., by solving the equations for the time evolution of the shape of nuclear surface of fssioning system. Here, we implicitly assume that a fissioning nucleus can be divided into 2 systems in thermal equilibrium at temperature *T*, namely, 1) heat bath consisting of nucleons and 2) the system of our interest, which consists of collective shape coordinates which are chosen to describe the change of nuclear shape during fission in a convenient way. These 2 systems are assumed to exchange energy via friction and random force. For the shape parametrization we use that of two-center model (TCM)[3] with 4 deformation parameters $q_\mu = \{ z_0/R_0, \delta_1, \delta_2, \alpha \}$, where z_0/R_0 refers to the distance between the centers of left and right Nilsson-type oscillator potentials with $R_0=1.2\ A^{1/3}$, R_0 being the radius of spherical nucleus with the mass number A. The parameters δ_i, where *i*=1,2, describe the deformation of the right and left fragment tips. The fourth parameter α is the mass asymmetry and the fifth parameter of TCM shape parametrization ε was kept constant, $\varepsilon = 0.35$, in all our calculations. This value of ε has been decided in a detailed analysis for fission of n+^{235}U system.

The first order stochastic differential equations (Langevin equations) for the time dependence of collective variables q_μ and the conjugated momenta p_μ are given as follows:

$$\frac{dq_\mu}{dt} = (m^{-1})_{\mu\nu} p_\nu , \tag{1}$$

$$\frac{dp_\mu}{dt} = -\frac{\partial F(q,T)}{\partial q_\mu} - \frac{1}{2} \frac{\partial m^{-1}_{\nu\sigma}}{\partial q_\mu} p_\nu p_\sigma - \gamma_{\mu\nu} m^{-1}_{\nu\sigma} p_\sigma + g_{\mu\nu} R_\nu (t), \tag{2}$$

where the sums over the repeated indices are assumed. In Eqs. (1) and (2), the $F(q, T)$ is the temperature dependent free energy of the system, and $\gamma_{\mu\nu}$ and $(m^{-1})_{\mu\nu}$ are the friction and inverse of mass tensors, $g_{\mu\nu}$ is the strength of random force and R_ν denotes a random force term corresponding to white noise.

The free energy $F(q, T)$ is calculated as the sum of liquid drop deformation energy and the temperature dependent shell correction $\delta F(q, T)$. The damping of shell correction $\delta F(q, T)$ with the excitation energy is described in detail in [4]. The single particle energies are calculated with the 2-center deformed Woods-Saxon potential [5,6] fitted to the aforementioned TCM shape parameterizations. It is to be noted the free energy is equal to potential energy at zero temperature. The main part of the shell correction consists of Strutinsky correction and BCS paring correction to the free energy.

The collective inertia tensor $m_{\mu\nu}$ is calculated within the Werner-Wheeler approximation [7] and for the friction tensor $\gamma_{\mu\nu}$ we used the wall-and-window formula[7,8]. The factors $g_{\mu\nu}$ are related to the temperature and friction tensor via the modified Einstein relation ,

$$g_{\mu\sigma} g_{\sigma\nu\mu\nu} = T^* \gamma \quad , \text{ with } T^* = \frac{\hbar\omega}{2} \coth \frac{\hbar\omega}{2T} , \tag{3}$$

where T^* is the effective temperature[9]. The parameter ω is the local frequency of collective motion [10]. The minimum of T^* is given by $\hbar\omega/2$. The temperature *T* in this context is related to the initial energy $E^*_{(in)}$ and the excitation energy E^* by,

$$E^* = E_{gs} + E^*_{(in)} - \frac{1}{2} m^{-1}_{\mu\nu} p_\mu p_\nu - V_{pot}(q, T=0) = aT^2, \tag{4}$$

where V_{pot} is the potential energy and a is the level density parameter. More details are given in our earlier publications [11-12].

Initially, the momenta p_μ are set to zero, and calculations are started from the ground state deformation. Such calculations are continued until the trajectories reach the "scission point", which was defined as the point in deformation space where the neck radius turns into zero

3 Results and discussion

Calculated fragment mass and TKE correlations for a set of selected actinide nuclei are shown in Fig. 1.

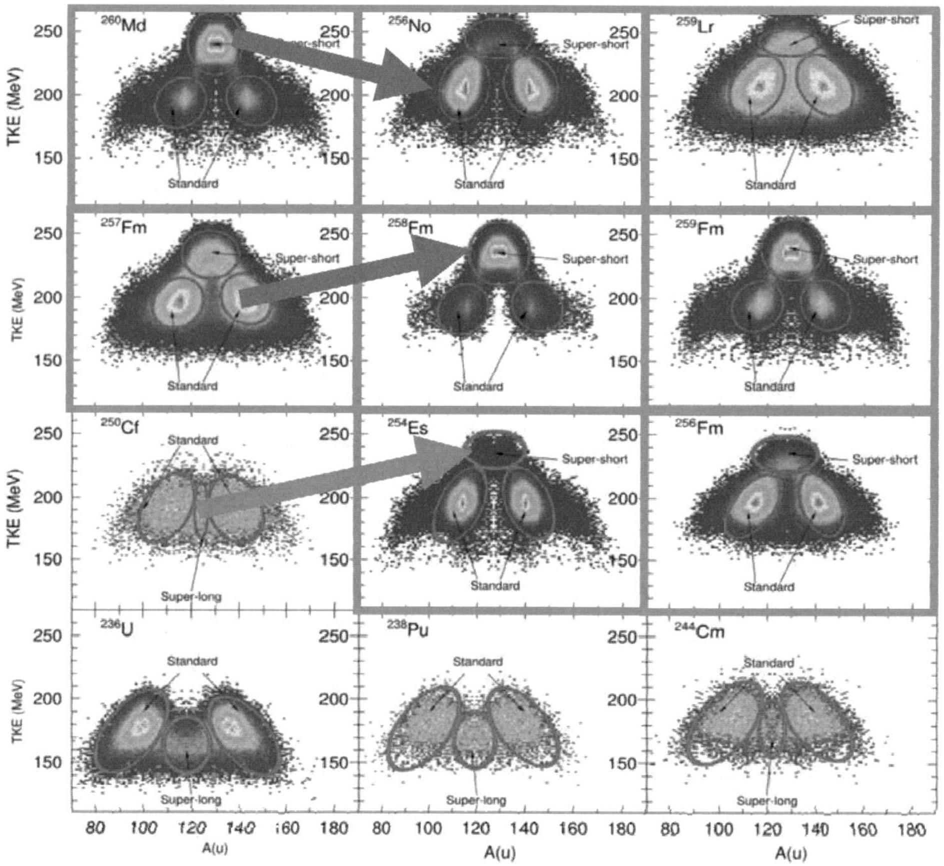

Fig. 1 Fragment mass and TKE correlations calculated by 4D Langevin model. The asymmetric (standard mode) and symmetric components are enclosed by red circles. The blue arrow shows a point of transition of the symmetric component from superlong to supershort mode, and the red arrows show transitions of the dominant mode from the standard to supershort mode and a reverse direction.

We can clearly observe a transition of the symmetric component, going from the superlong to supershort modes between ^{250}Cf and ^{254}Es, and the symmetric component stays at the supershort mode for heavier systems. On the contrary, the dominant mode makes a transition in between ^{257}Fm to ^{258}Fm from asymmetric to symmetric mode, (mass distribution changes from 2-peaked structure to a single-peaked one) while it transits back to the asymmetric mode in between ^{260}Md to ^{256}No (transition of mass distribution from single-peak to 2-peaked distribution).

We can calculate the average TKE of the symmetric and asymmetric components separately, and they are plotted as a function of $Z^2/A^{1/3}$ of the fissioning nucleus in Fig. 2. It can be seen from this figure that the average TKE for the asymmetric component, "Calc. TKE_{asy}", increases monotonically, and it is in very good agreement with systematics proposed by Unik [13] and Viola [14]. On the other hand, the average TKE for the symmetric component makes a sudden jump at Cf to Fm as indicated by the blue arrow. Therefore, there is a clear transition of the fission mechanisms here. The black dots in Fig. 2 denote position of the dominant mode. It is on the standard mode for nuclei having smaller $Z^2/A^{1/3}$, showing that the mass distribution has 2 peaks (since it is in the asymmetric mode), then makes a sudden jump to the symmetric component in the middle of the Fm isotopes so the mass distribution turns into a single-peaked structure, then suddenly transits back to the asymmetric mode as the red arrows indicated. By the combination of these 2 transitions, we can explain very well the sudden change of the mass distribution from ^{257}Fm to ^{258}Fm, and systematics and anomaly in the average TKE of fission fragments.

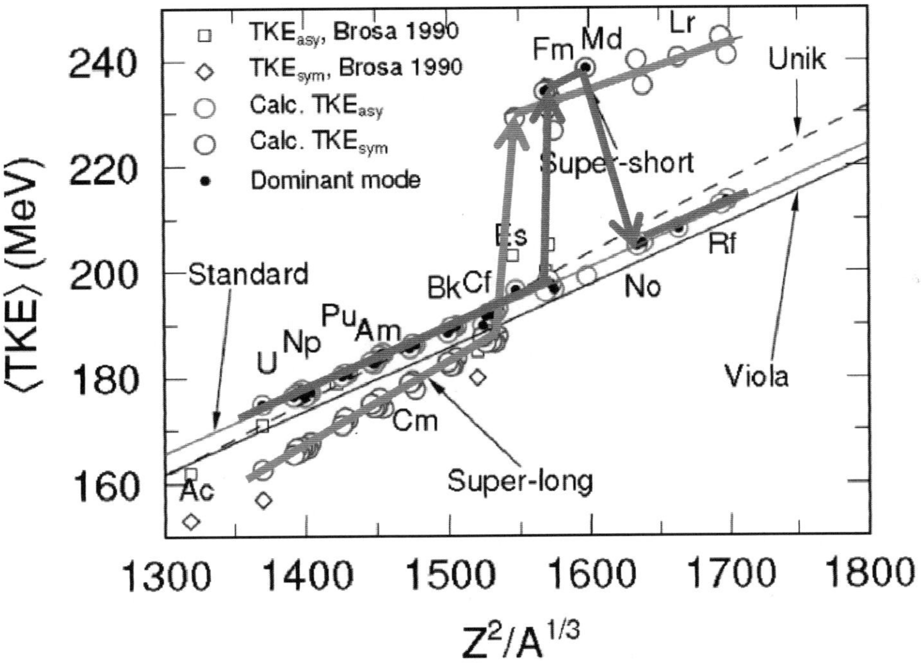

Fig. 2 Average total kinetic energy (<TKE>) of the symmetric and asymmetric components as a function of $Z^2/A^{1/3}$ of the fissioning nuclei. The calculated values are denoted as "Calc. TKE_{asy}" for the asymmetric component, and "Calc. TKE_{sym}" for the symmetric component. The dominant mode is indicated by the black dots. The square and diamond symbols are experimental data reported in Brosa's paper 15].

Figure 3 shows mass distributions (upper panels) and mass-quadrupole moment (bottom panels) for ^{256}Fm and ^{258}Fm. It is well known that the mass distribution becomes from 2 peaked structure typical for actinides to a sharply single-peaked structure here. We think we know the reason. If we see the quadrupole moment of ^{258}Fm, the average value at each fragment mass number, shown by the black histogram, touches down to Q_{20}=0 at A=132. It clearly shows that the fission mechanisms is dominated by the spherical magicity of ^{132}Sn, and, since the mass number is close enough to twice 132, the compliment fragment should have a very similar mass number, so both the light and heavy fragments form a common single peak. On the contrary, the average Q_{20} dese not touches down to 0 for ^{256}Fm, but stays finite. It indicates that the fission is affected by the deformed magicity around A=142-144[16]. Since the mass number of the heavy fragments is large enough, the compliment light fragments have mass numbers which are sufficiently small to form another peak, giving rise to the 2-peaked structure. We carried out similar analysis for Q_{30}, but we will omit discussion on it here.

Conclusions

We have investigated the mass-TKE correlation of fission fragments by 4D Langevin model. The calculated data were decomposed into symmetric and asymmetric components and we have seen how they behave when fissioning system changes. We have observed that there is a clear transition of 1) the symmetric component from the superlong to supershort mode, and 2) dominant mode from asymmetric to symmetric, then to asymmetric component. We have shown that combination of these 2 transitions can account for the systematic and anomalous trends in mass and TKE distributions in the region of actinides. Furthermore, the mass-quadrupole deformation can explain why the sudden transition from the 2-peak structure to a single-peak structure in mass distribution between ^{256}Fm (actually ^{257}Fm) and ^{258}Fm; fission in nuclei lighter than ^{257}Fm is affected by deformed magicity at A=142-144 as suggested by Scamps and Simenel, while that of ^{258}Fm is strongly affected by spherical magicity of ^{132}Sn.

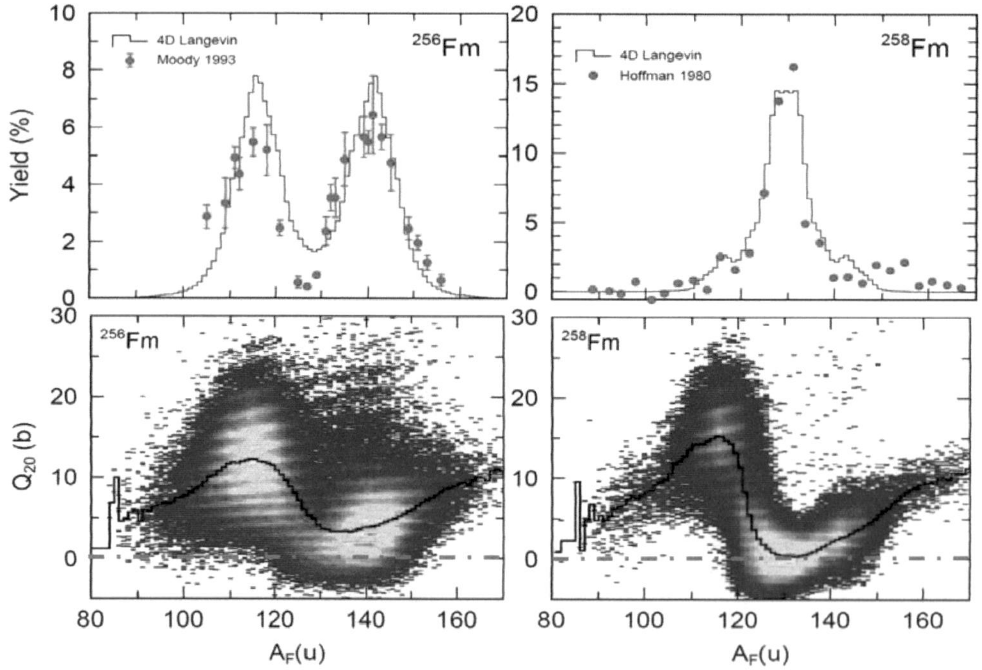

Figure 3 shows mass distributions (upper panels) and mass-quadrupole moment (bottom panels) for ^{256}Fm and ^{258}Fm.

References

1. M.D.Usang, F.A.Ivanyuk, C.Ishizuka and S.Chiba, Scientific Reports **9**, 1525(2019)
2. Y.Abe, S.Ayik, P.-G.Reinhard and E. Suraud, Phys. Rep. 275, 49-196 (1996).
3. J.Maruhn and W.Greiner, Zeit. f. Phys. 251, 431-457 (1972).
4. F.A.Ivanyuk, C.Ishizuka, M.D.Usang and S.Chiba, Phys. Rev. C 97, 054331 (2018).
5. V.V.Pashkevich, Nucl. Phys. A 169, 275-293 (1971).
6. V.V.Pashkevich, Nucl. Phys. A 477, 1-17 (1988).
7. K.T.R.Davies, A.J.Sierk and R.Nix, Phys. Rev. C 13, 2385-2403 (1976).
8. J.Blocki et al Ann. Phys. 113, 330-381 (1978).
9. A.J.Sierk and J.R.Nix, Phys. Rev. C 21, 982-987 (1980).
10. H.Hofmann and D.Kiderlen, Int. Jour. Mod. Phys. E 7, 243-274 (1998).
11. M.D.Usang, F.A.Ivanyuk, C.Ishizuka and S.Chiba, Phys. Rev. C 94, 044602 (2016).
12. C.Ishizuka, M.D.Usang, F.A.Ivanyuk, J.A.Maruhn, K.Nishio and S.Chiba, Phys. Rev. C 96, 064616 (2017).
13. J.P.Unik, and J.E.Gindler, "Critical review of the energy released in nuclear fission", ANL-7748. Tech. Rep., Argonne National Lab., Ill.(1971).
14. V.E.Viola, K.Kwiatkowski, and M.Walker, Phys. Rev. C 31, 1550 (1985).
15. U.Brosa, S.Grossmann and A.Mueller, Phys. Rep. **197**, 167-262(1990)
16. G.Scamps and C.Simenel, Nature 546, 382-385(2018)

Description of the Fission Process: Nuclear Models for Fission Dynamics

M. Verriere[1,*], *M.R.* Mumpower[1,**], *T.* Kawano[1,], and *N.* Schunck[2,]

[1]Los Alamos National Laboratory, Los Alamos, NM 87545, USA
[2]Nuclear and Chemical Sciences Division, Lawrence Livermore National Laboratory, Livermore, CA 94551, USA

Abstract. Nuclear fission is the splitting of a heavy nucleus into two or more fragments, a process that releases a substantial amount of energy. It is ubiquitous in modern applications, critical for national security, energy generation and reactor safeguards. Fission also plays an important role in understanding the astrophysical formation of elements in the universe. Eighty years after the discovery of the fission process, its theoretical understanding from first principles remains a great challenge. In this paper, we present promising new approaches to make more accurate predictions of fission observables.

1 Introduction

On the one hand, a theoretical understanding of the nuclear fission process is essential for compiling nuclear data associated with reactions inaccessible to experiment. It is also of first importance to unravel the origin of elements in the solar system as well as in the universe, through the description of the fission decay following the rapid neutron capture process, involving many neutron-rich fissioning systems (FS) [1, 2]. On the other hand, several experimental campaigns have been accumulating a number of observables associated with different steps of the fission process with smaller and smaller uncertainties for an always wider range of nuclei [3–7], which could be used to constrain the theory. However, results obtained from predictive theories of the fission process are rather far from experimental data, or only able to predict a small subset of fission observables, making the testing of the different theories a difficult task.

In this work, we focus on developing an approach to predict the probability distribution of the fission fragments mass and charge Y(Z,A) before prompt emission. In Sect. 2, we sumarize the main characteristics of our approach. In Sect. 3, we compare predictions of initial fission fragment distributions for Uranium isotopes agains experimental data and evaluations.

2 Theoretical Approach

Our approach to the description of the fission process can be decomposed into three steps. The potential energy surface (PES) associated with the FS of interest is first obtained from

*e-mail: verriere@lanl.gov
**e-mail: mumpower@lanl.gov

semi-classical considerations (see 2.1). The PES gives the potential energy of the FS as a function of its shape, which is used in the second step as an input for estimating the density of probability to populate a given scission configuration. The last step consists in the determination of the fragments mass and charge probabilities (see 2.2).

2.1 Potential Energy Surfaces

Here, the deformations of the fissioning system are defined through the five-dimensional Matched-Quadratic-Surface parameterization of the geometric shape associated with the deformation (by setting the center of mass of the shape to be at the origin and its volume to be constant) [8, 10]. In the macroscopic-microscopic method, the potential energy E of the fissioning system at a given deformation is written as the sum of a macroscopic term E_{mac}, which corresponds to the smooth part (with respect to particle number) of the PES, and a microscopic corrective term $\Delta E_{\text{shell}} + \Delta E_{\text{pair}}$ arising from shell and pairing effects in the FS,

$$E = E_{\text{mac}} + \Delta E_{\text{shell}} + \Delta E_{\text{pair}}. \tag{1}$$

The first term of the right-hand side of (1) is calculated by assuming that the FS is a deformed drop of a charged nuclear fluid assuming a finite-range interaction between the volume elements of the drop, in the same way as developed in [9], but using the parameterization of the Liquid-Drop Energy presented in [11].

The second term ΔE_{shell} is calculated with the same average nuclear potential as in [9]. The solutions to the Schrödinger equation are expanded in an axial harmonic oscillator (AHO) basis, but with three major improvements:

1. the two parameters of the AHO basis (the frequencies of the oscillator along the x- and z-axes) are optimized according to a variational principle;
2. the spurious effects of continuum states are reduced by using the improved Strutinsky method presented in [12], instead of the standard one;
3. the last point allowed us to increase the number of shells in the AHO basis from twelve to twenty.

Finally, the last term ΔE_{pair} is calculated by solving the BCS equation with the Lipkin-Nogami (LN) correction for a seniority pairing force. We use a valence space that contains all canonical states located within ±5 MeV around the Fermi level. We solve the LN equations numerically using the method developed by J.R. Sharma and P. Gupta in [13]. This numerical method requires the inversion of the Jacobian matrix associated with the LN equations at each iteration. To improve the stability and the performances of the code, this is done analytically, only inverting 2×2 and 3×3 matrices. This implementation allows solving the LN equations with an error $\varepsilon \approx 10^{-10}$ in a completely negligible time compared to the one required to calculate ΔE_{shell}.

2.2 Scission configurations and fragmentation yields

In order to estimate the relative probabilities of scission configurations, we have developed a novel method that directly computes the actual density of probability instead of sampling it through the calculation of multiple trajectories, as in the standard implementation of a random walk. This new method is implemented in a code call Deterministic-Probabilistic Solver (DPS) and does not require to "tilt" the PES in order to pass potential barriers (see discussion on bias potentials in [14]). By construction, each time iteration in the DPS framework is longer than with the standard implementation since all possible trajectories are considered at once. However, as shown in Figure 1, the relative fragmentation probability computed in the DPS converges extremely fast, with an error below 0.1% after only 10,000 (^{234,236,238}U) and 20,000 (^{240}U) time iterations. With the standard implementation of the random walk, a calculation would require of the order of $\sim$ 200,000 time iterations to reach scission and repeat $\sim$ 50,000 time the full calculation to obtain the same precision.

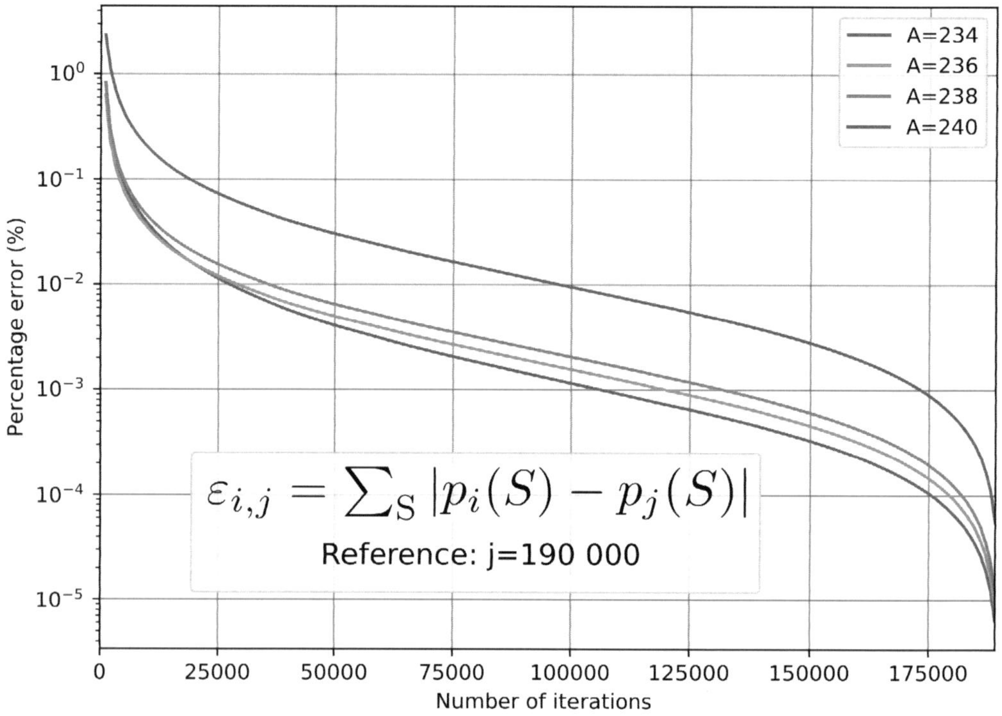

Figure 1. Relative error in the fragment distribution within DPS as a function of the number of time iterations (Z=92).

From the population probability $P(S)$ of a given scission configuration S obtained with DPS, we calculate the correlated mass and charge yields Y(Z,A)

$$Y(Z,A) = \sum_{S} P(S) \times P_S(Z,A), \tag{2}$$

where $P_S(Z,A)$ is the probability to measure the mass and charge distribution of the fragments associated with the scission configuration S. This quantity is estimated by using the generalization of the particle-number projection technique developed in [15], adapted to fission purposes in the case of TDHF+BCS in [16], and for collective models in [17]. Note that we do not account for possible correlations in the fragment observables between different points of the PES. Uncorrelated charge, Y(Z), and mass, Y(A), yields are obtained by integration of Y(Z,A) over A and Z, respectively.

3 Application to Thermal Fission of Uranium Isotopes

The first reactions we have considered in this study are thermal fission for the isotopic chain of even-even isotopes of Uranium from $A = 236$ to $A = 242$. We applied the approach presented in Sec. 2 using, as emphasized above, a more complete harmonic oscillator basis basis with 20 shells instead of 12 as in previous studies such as [10, 18, 19]. In Sect. 3.1, we first show the uncorrelated mass, Y(A), and charge, Y(Z), fragmentation probabilities before prompt emission. We then show in Sect. 3.2 that our approach is also capable of estimating the fully correlated charge and mass fragmentation probabilities Y(Z,A) for the reaction ^{235}U(n,f).

3.1 Uncorrelated pre-neutrons mass and charge yields

The pre-neutron emission mass and charge yields we have obtained for the reactions ^{235,237}U(n,f) are shown in figure 2. With the exception of the case $Z_{\text{frag}} = 36$, we see that

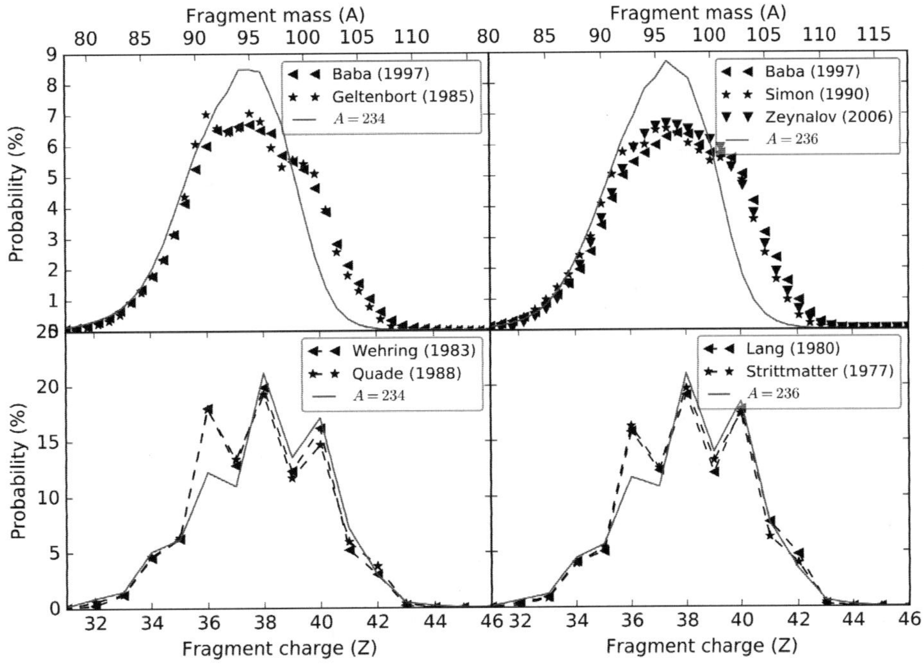

Figure 2. Mass (top) and charge(bottom) fragmentation distributions of the light fragment before prompt emission obtained for the reactions ^{235}U(n,f) (left) and ^{237}U(n,f) (right) at thermal neutron energy compared with experimental data from EXFOR.

the fragment charge yields are extremely well reproduced. The mass yields are in qualitatively good agreement with experimental data (that are all corrected to take into account prompt neutron emission). However, the average mass of the yields is not well reproduced. According to our analysis, this discrepancy probably comes from the way we estimate the neck position along the z-axis: we calculate it from the total 1-body local density (neutron and protons), while we should consider separately the neutron and the proton local density because of the absence of interaction between both isospins in our formalism. Also, the width of the mass distribution does not match the experimental data. Even though a deeper analysis is necessary to fully understand the origin of this discrepancy, we can formulate two hypotheses: i) the mass resolution of the *experimental* data is not included in our calculation and could lead to an increase of the FWHM of the yields of $\Delta A_{\text{frag}} \sim 2.5$, ii) the collective space is too small, in particular along the dimensions corresponding to the deformations of the fragments (ε_L and ε_R in [10]): this introduces a bias in the probability distribution, which cannot probe all possible fragmentations.

We have also used our approach to predict Y(A) and Y(Z) for the reactions ^{239,241}U(n,f), and our results are presented in figure 3. The charge yields obtained for the reaction ^{239}U(n,f) matches quantitatively well the evaluation from JENDL.

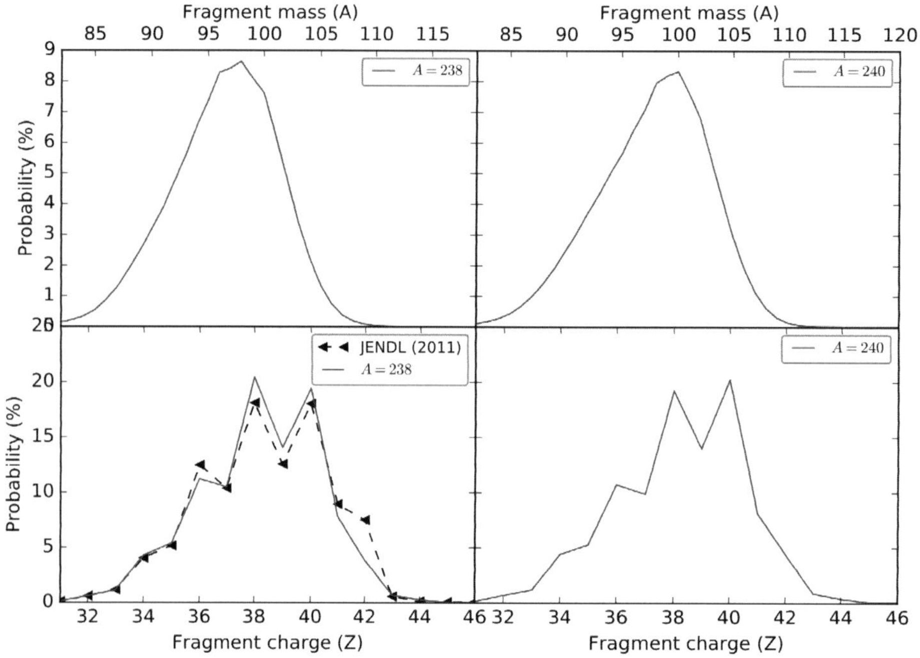

Figure 3. Same as figure 2 for the reactions ^{239}U(n,f) (left) and ^{241}U(n,f) (right).

3.2 Correlated pre-neutrons mass and charge yields

The primary output of our method is the correlated yields in mass and charge Y(Z,A) given by Eq. (2). The figures 4 and 5 show these quantities for the reaction ^{233}U(n,f) and ^{235}U(n,f), respectively, compared with independent yields taken from experimental data.

Both results correspond to asymmetric fission. The misalignment between the probability contours associated with the light and heavy fragment, called the charge polarization of the fragment distributions, is well reproduced for the first time in this type of approach. However, the probability contours are wider than the experimental ones. A possible origin of this mismatch between our calculations and the experimental data can be missing correlations between neutrons and protons in our calculation, where such correlations are only taken into account at the dynamical level (Random Walk) without being included in the calculation of the PES itself.

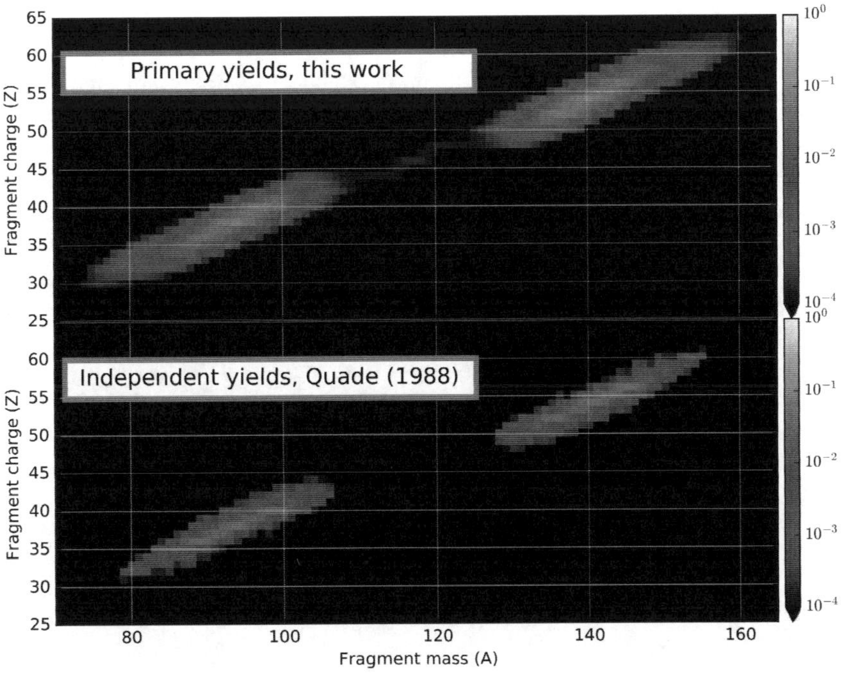

Figure 4. Correlated yields in mass and charge for the reaction ^{233}U(n,f) obtained with our work before prompt emissions (upper panel) and from experimental data after prompt neutron emissions (lower panel).

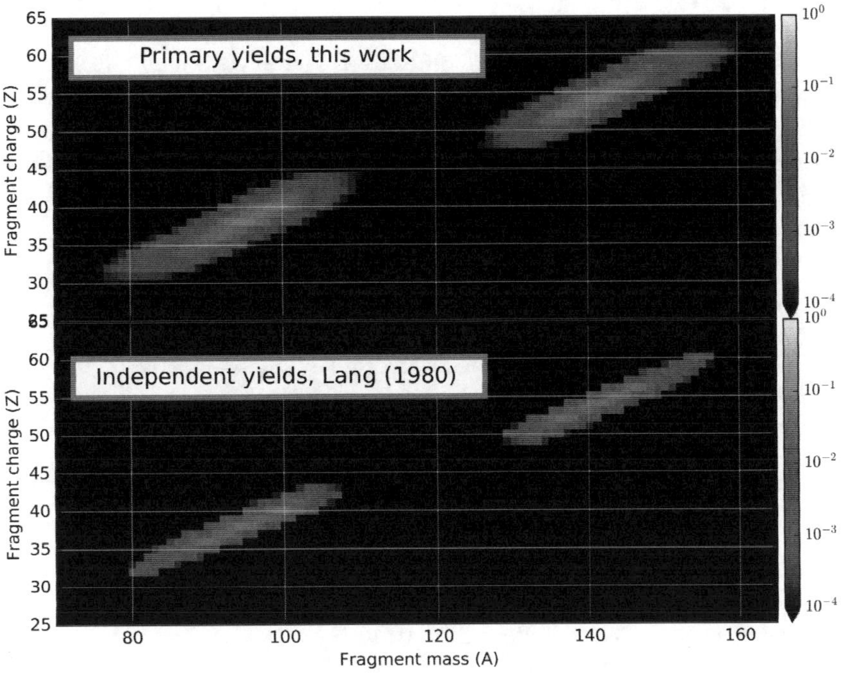

Figure 5. Same as figure 4 for the reaction ^{235}U(n,f).

4 Conclusion

In this work, we have extended the Los Alamos method to estimate the fission yields by including more microscopic correlations in the calculation. To do so, we have used an improved Strutinsky method to reduce the spurious effect of continuum states. This method allowed us to use 20 harmonic oscillator shells instead of 12 in previous calculations. We have also incorporated a new method based on particle number projection to estimate the mass and charge fragmentation probabilities associated with a given scission configuration in the PES. The charge yields obtained with our method are in good quantitative agreement with experimental data, reproducing the odd-even staggering for the first time in this type of approach. We are also able to predict the correlated mass and charge yields Y(Z,A) with a good reproduction of charge polarization effects of the yields.

This work was performed at Los Alamos National Laboratory under the auspices of the National Nuclear Security Administration of the U.S. Department of Energy at Los Alamos National Laboratory under Contract No. 89233218CNA000001. Support for this work was provided through the Fission In R-process Elements (FIRE) Topical Collaboration in Nuclear Theory of the U.S. Department of Energy. Part of this work was performed under the auspices of the U.S. Department of Energy by Lawrence Livermore National Laboratory under Contract DE-AC52-07NA27344

References

[1] G. Martínez-Pinedo, D. Mocelj, N.T. Zinner et al., Progress in Particle and Nuclear Physics **59 1**, 199-205 (2007)
[2] N. Vassh, R. Vogt, R. Surman, J. Randrup et al., J. Phys. G: Nucl. Part. Phys. **46** 065202 (2019)
[3] K.-H. Schmidt, J. Benlliure, A.R. Junghans, Nuclear Physics A **693 1-2**, 169-189 (2001)
[4] B. Jurado, C. Schmitt, K.-H. Schmidt et al., Phys. Rev. Lett. **93**, 072501 (2004)
[5] G. Boutoux, G. Bélier, A. Chatillon, A. Ebran, T. Gorbinet, B. Laurent, J.-F. Martin, E. Pellereau, J. Taieb et al., Physics Procedia **47**, 166-171 (2013)
[6] R.C. Haight et al., Nuclear Data Sheets **119**, 205-208 (2014)
[7] K. Meierbachtol, F. Tovesson, D. Shields et al., Nuclear Instruments and Methods in Physics Research Section A: Accelerators, Spectrometers, Detectors and Associated Equipment **788**, 59-66 (2015)
[8] J.R. Nix, Nuclear Physics A **130 (2)**, 241-292 (1969)
[9] P. Möller, A.J. Sierk, T. Ichikawa, H. Sagawa, Atomic Data and Nuclear Data Tables **109-110**, 1-204 (2016)
[10] P. Möller, A.J. Sierk, T. Ichikawa, A. Iwamoto, R. Bengtsson, H. Uhrenholt, S. Åberg, Phys. Rev. C **79 (6)**, 064304 (2009)
[11] P. Moller, A.J. Sierk, A. Iwamoto, Phys. Rev. Lett **92 (7)**, 072501 (2004)
[12] N. Tajima, Y.R. Shimizu, S. Takahara, Phys. Rev. C **82 (3)**, 034316 (2010)
[13] J.R. Sharma, P. Gupta, Computers & Mathematics with Applications **67 (3)**, 591-601 (2014)
[14] J. Randrup, P. Möller, and A. J. Sierk, Physical Review **C** 84, 034613 (2011)
[15] C. Simenel, Phys. Rev. Lett. **105**, 192701 (2010)
[16] G. Scamps, D. Lacroix, Phys. Rev. C **87**, 014605 (2013)
[17] M. Verriere, N. Schunck, T. Kawano, Phys. Rev. C **100 (3)**, 034615 (2019)
[18] P. Moller, J. Randrup, A.J. Sierk, Phys. Rev. C **85**, 024306 (2012)
[19] P. Moller, J. Randrup, Phys. Rev. C **91**, 044316 (2015)

Following Fission Products in Explosive Astrophysical Environments

T. M. Sprouse[1,2], *M. R.* Mumpower[2,3], and *R.* Surman[1]

[1]Department of Physics, University of Notre Dame, Notre Dame, Indiana, USA
[2]Theoretical Division, Los Alamos National Laboratory, Los Alamos, New Mexico, USA
[3]Center for Theoretical Astrophysics, Los Alamos National Laboratory, Los Alamos, New Mexico, USA

Abstract. The astrophysical process by which the heaviest elements are formed in the universe is known as the rapid neutron capture process, or *r* process, of nucleosynthesis. The *r* process is characterized by the neutron capture and β^- decay of short-lived, neutron-rich atomic nuclei; in suitably extreme environments, nuclear fission can also play a major role in determining the ensuing nucleosynthesis. In this work, we present the application of our recently-developed *nucleosynthesis tracing framework* to precisely quantify the impact that neutron-induced and β^--delayed fission processes have in *r*-process environments that produce fissioning nuclei.

1 Introduction

The rapid neutron capture (*r*) process of nucleosynthesis is thought to be the astrophysical process by which the heaviest elements are formed in the universe. While the general mechanism has been proposed since the 1950s [1], a complete understanding of its mysteries has, to date, remained elusive. Considerable challenges exist across a range of disciplines relevant to the *r* process, including in observational astronomy, computational astrophysics, and computational and experimental nuclear physics [2].

While significant progress is underway in each of these fields to further our understanding of the *r* process (see, e.g., [3] for a recent review), this work focuses specifically on the problem of modeling and analyzing the competing processes that occur during nucleosynthesis. In astrophysical environments conducive to nucleosynthesis, a system of atomic nuclei may undergo any number of nuclear transmutation processes; through these processes, the abundance of each nuclear species evolves over time. The *r* process is an especially complex example of nucleosynthesis which involves thousands of nuclear species connected by tens of thousands of transmuation processes. Nevertheless, numerical techniques exist to effectively model how the nuclear abundances evolve over time in astrophysical *r*-process environments [4–9], and a great deal has been learned about the *r* process through their use.

However, *r*-process nucleosynthesis (and nucleosynthesis in general) is a fundamentally dynamical and non-linear problem. Because of this, it is extremely difficult to establish the relationship between a particular nucleus or transmutation process to the nucleosynthetic system as a whole. Approximate statements may be made, for example, by analyzing how fast or how often the transmutation process occurs, as in [10–12], but these approaches lack consideration for how the process actually *alters* any nucleosynthesis that takes place after it occurs.

In this work, we present recent results that attempt to address this challenge. In Sect. 2, we introduce our recently-developed *nucleosynthesis tracing framework* which allows for a precise and quantitative description of the role nuclear transmutation processes have in nucleosynthetic systems. In Sect. 3, we demonstrate several immediate applications of this framework to characterize how fission products participate in a fission-cycling *r* process.

2 Nucleosynthesis Tracing

We have developed a nucleosynthesis tracing framework that extends the function of conventional nuclear reaction networks to enable a quantitative description of individual nuclear reaction, decay, and fission transmutation processes, or collections thereof, as they participate in the broader context of nucleosynthetic events. We begin by identifying all nuclei in a calculation as belonging to either a *traced* or *untraced* population. At the beginning of each calculation, all nuclei are assigned to the untraced population. Additionally, we allow each of the reaction, decay, and fission processes occurring during nucleosynthesis to be one of three types, based on how the products of each are assigned to the traced and untraced populations. In particular, the products of each process may be

- mapped exclusively to the traced population,
- mapped to preserve the identity of the reactant(s), or
- mapped exclusively to the untraced population,

as illustrated schematically in figure 1. By carefully assigning each process to one of these three types, one may effectively model all subsequent nucleosynthesis in which fission products participate, among other exciting applications. For a more thorough description of the tracing framework, as well as some comments on its implementation in the nuclear reaction network code Portable Routines for Integrated nucleoSynthesis Modeling (PRISM), see [13].

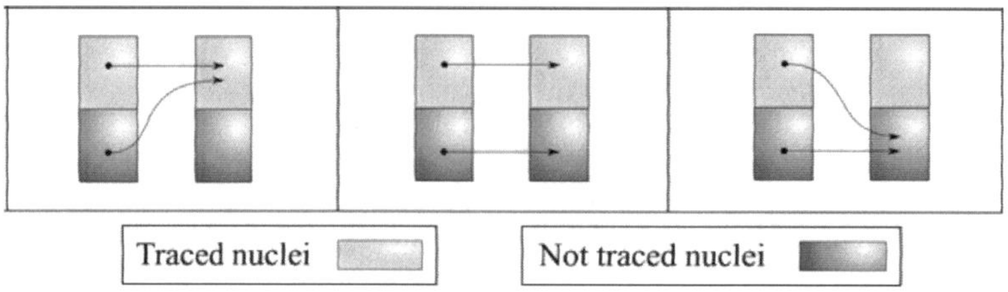

Figure 1. Schematic overview of the nucleosynthesis tracing framework. All nuclei belong to either a traced population (top red boxes) or an untraced population (bottom blue boxes). Every nuclear reaction, decay, and fission channel is cast into one of three types based on how it maps its products into the traced and untraced populations. The first type maps product(s) exclusively to the traced population (left panel). The second type preserves the traced and untraced populations between the reactant(s) and product(s) (center panel). The last type maps product(s) exclusively to the untraced population (right panel).

3 Results

We present several immediate applications of the nucleosynthesis tracing framework described in Sect. 2 to describe the interplay between nuclear fission and *r*-process nucleosynthesis. We begin with astrophysical conditions taken from a hydrodynamical simulation of material ejected during a neutron star merger [14]. For these conditions, fission is an especially active process, such that nearly all nuclei have participated in fission at least once during nucleosynthesis. All relevant reaction, decay, and fission rates, as well as β-delayed neutron emission probabilities, are the same as used in [11]. Fission yields are taken from calculations based on the finite-range liquid-drop model [15].

3.1 Neutron-induced Fission Products

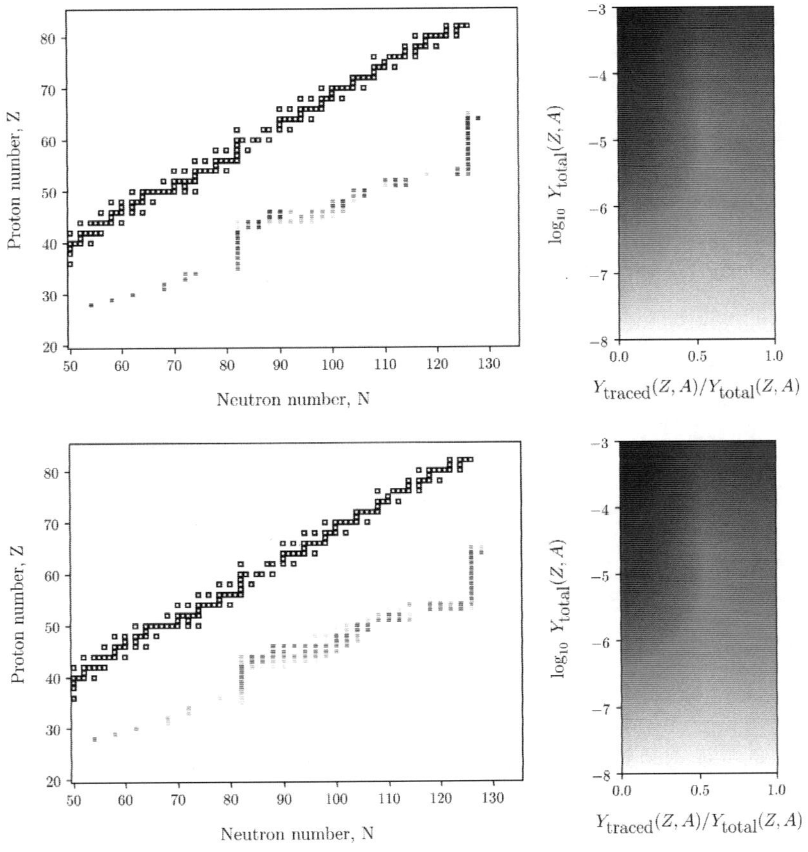

Figure 2. Neutron-induced fission product contributions to nuclear abundances at the onset of fission during nucleosynthesis (top panel) and after an additional $\sim$ 0.2 seconds of nucleosynthesis has taken place (bottom panel). Total abundances ($Y_{\rm total}$) of each nucleus is given by the luminosity of the colored squares, and their hue encodes the relative fraction of traced nuclei to the total ($Y_{\rm traced}/Y_{\rm total}$). In the top panel, nuclei with mass number $A \sim 130$ are beginning to be dominated by fission product contributions. In the bottom panel, effectively all nuclei with $A > 120$ have been involved with neutron-induced fission.

As a first example, we perform a nucleosynthesis calculation in which all neutron-induced fission products are mapped into the traced population. In this way, we quantify the relative fraction of each nuclear abundance that has a history involving a neutron-induced fission event, at each timestep in the calculation. In figure 2, we plot this fraction at two points in time during nucleosynthesis. In the top panel, we consider the approximate onset of neutron-induced fission, in which nuclei in the neighborhood of $A \sim 130$ begin to become populated with the products of fissioning $A > 250$ nuclei. These fission products will capture additional neutrons and proceed along the r-process pathway to form heavier nuclei. After an additional 0.2 seconds of nucleosynthesis has elapsed, sufficient neutron-induced fission has taken place that effectively all heavy nuclei have a history involving neutron-induced fission, as shown in the bottom panel of figure 2.

3.2 Fission Products of Individual Nuclei

We may further refine the approach taken in Sect. 3.1 to study the fission products of individual nuclides fissioning by distinct fission channels. In figure 3, we present such an analysis applied to the neutron-induced fission of Plutonium 295 and the β^--delayed fission of Berkelium 268. Plutonium 295 is produced along with many other very neutron-rich actinides at early stages of the r process, when the abundance of free neutrons is high. Therefore, its fission products undergo substantial reprocessing by subsequent neutron capture, photodissociation, and β^--decay processes. As such, these fission product yields bear minimal resemblance to their distribution at the conclusion of nucleosynthesis; the fission product yield distribution is 'washed away' by other reaction and decay processes.

This contrasts with the β^--delayed fission of Berkelium 268. This nucleus fissions at a relatively late stage of the r process, as neutron capture is winding down. Thus its fission products are predominantly restricted to β^- decays which mostly preserve their isotopic distribution, as shown in figure 3. In this case, the fission product yields are critical for determining their contribution to final nucleosynthetic yields.

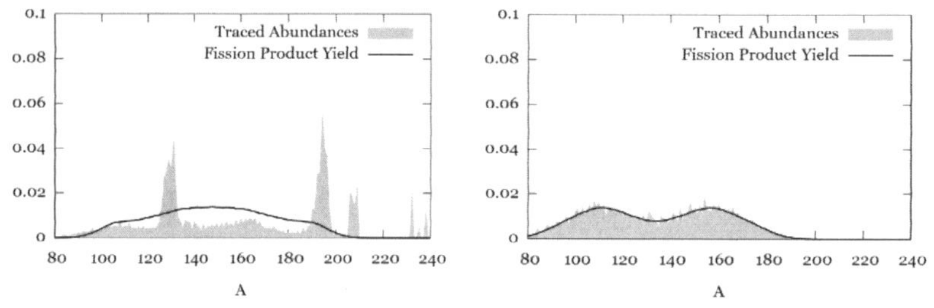

Figure 3. Comparison of mass distribution of fission products immediately after fission occurs (blue line) and after they participate in all subsequent r process nucleosynthesis (light-blue filled curve). The left panel corresponds to the neutron-induced fission of Plutonium 295; this nucleus fissions relatively early during nucleosynthesis, so its fission products undergo a significant amount of subsequent nucleosynthesis. The right panel corresponds to the β^--delayed fission of Berkelium 268; in this case, fission occurs at a much later time during nucleosynthesis, and the products' contribution to nucleosynthesis is closely tied to its product yield distribution.

4 Conclusions

Nuclear fission may play an important role during the formation of the heaviest elements via the astrophysical r process [16]; through the use of our nucleosynthesis tracing framework, we can drastically improve our ability to quantitatively describe this role. In this work, we have demonstrated some of the analyses that this technique enables. In particular, we show that it is possible to follow fission products generated during the *r* process throughout any follow-up nucleosynthesis in which they participate; this approach can be applied to the fission of all nuclei via a particular channel, as in Sec. 3.1, or even for the fission of individual nuclei, as in Sec. 3.2.

5 Acknowledgements

This work was supported in part by the US Department of Energy under Contract No. DE-FG02-95-ER40934, the topical collaboration Fission In R-process Elements (FIRE) Contract No. DE-AC52-07NA27344, and SciDAC Contract No. DE-SC0018232. MM was supported by the US Department of Energy through Los Alamos National Laboratory and by the Laboratory Directed Research and Development program of Los Alamos National Laboratory under project number 20190021DR. Los Alamos National Laboratory is operated by Triad National Security, LLC, for the National Nuclear Security Administration of U.S. Department of Energy (Contract No. 89233218CNA000001). T.S. was supported in part by the Los Alamos National Laboratory Center for Space and Earth Science, which is funded by its Laboratory Directed Research and Development program under project number 20180475DR.

References

[1] E.M. Burbidge, G.R. Burbidge, W.A. Fowler, F. Hoyle, Reviews of Modern Physics **29**, 547 (1957)

[2] C.J. Horowitz, A. Arcones, B. Côté, I. Dillmann, W. Nazarewicz, I.U. Roederer, H. Schatz, A. Aprahamian, D. Atanasov, A. Bauswein et al., Journal of Physics G Nuclear Physics **46**, 083001 (2019), 1805.04637

[3] T. Kajino, W. Aoki, A.B. Balantekin, R. Diehl, M.A. Famiano, G.J. Mathews, Progress in Particle and Nuclear Physics **107**, 109 (2019), 1906.05002

[4] W.D. Arnett, J.W. Truran, The Astrophysical Journal **157**, 339 (1969)

[5] S.E. Woosley, W.D. Arnett, D.D. Clayton, The Astrophysical Journal Supplement Series **26**, 231 (1973)

[6] F.X. Timmes, The Astrophysical Journal Supplement Series **124**, 241 (1999)

[7] W.R. Hix, F.K. Thielemann, Journal of Computational and Applied Mathematics **109**, 321 (1999), astro-ph/9906478

[8] W.R. Hix, B.S. Meyer, Nuclear Physics A **777**, 188 (2006), astro-ph/0509698

[9] J. Lippuner, L.F. Roberts, The Astrophysical Journal Supplement Series **233**, 18 (2017), 1706.06198

[10] S. Wanajo, The Astrophysical Journal Letters **666**, L77 (2007), 0706.4360

[11] M.R. Mumpower, T. Kawano, T.M. Sprouse, N. Vassh, E.M. Holmbeck, R. Surman, P. Möller, The Astrophysical Journal **869**, 14 (2018), 1802.04398

[12] N. Vassh, R. Vogt, R. Surman, J. Randrup, T.M. Sprouse, M.R. Mumpower, P. Jaffke, D. Shaw, E.M. Holmbeck, Y. Zhu et al., Journal of Physics G Nuclear Physics **46**, 065202 (2019), 1810.08133

[13] T.M. Sprouse, M.R. Mumpower, R. Surman, In Prep. (2020)

[14] T. Piran, E. Nakar, S. Rosswog, Monthly Notices of the Royal Astronomical Society **430**, 2121 (2013), 1204.6242

[15] M.R. Mumpower, P. Jaffke, M. Verriere, J. Randrup, arXiv e-prints arXiv:1911.06344 (2019), 1911.06344

[16] M. Eichler, A. Arcones, A. Kelic, O. Korobkin, K. Langanke, T. Marketin, G. Martinez-Pinedo, I. Panov, T. Rauscher, S. Rosswog et al., The Astrophysical Journal **808**, 30 (2015)

Probing the fission properties of neutron-rich actinides with the astrophysical *r* process

Nicole Vassh[1,6,*], *Matthew* Mumpower[2,6], *Trevor* Sprouse[1,3], *Rebecca* Surman[1,6], and *Ramona* Vogt[4,5]

[1]University of Notre Dame, Notre Dame, Indiana 46556, USA
[2]Theoretical Division, Los Alamos National Lab, Los Alamos, NM, 87545, USA
[3]Los Alamos Center for Space and Earth Science Student Fellow, Los Alamos National Lab, Los Alamos, NM, 87545, USA
[4]Nuclear and Chemical Science Division, Lawrence Livermore National Laboratory, Livermore, CA 94551, USA
[5]Department of Physics, University of California, Davis, CA 95616, USA
[6]Joint Institute for Nuclear Astrophysics - Center for the Evolution of the Elements, USA

Abstract. We review recent work examining the influence of fission in rapid neutron capture (*r*-process) nucleosynthesis which can take place in astrophysical environments. We briefly discuss the impact of uncertain fission barriers and fission rates on the population of heavy actinide species. We demonstrate the influence of the fission fragment distributions for neutron-rich nuclei and discuss currently available treatments, including recent macroscopic-microscopic calculations. We conclude by comparing our nucleosynthesis results directly with stellar data for metal-poor stars rich in *r*-process elements to consider whether fission plays a role in the so-called 'universality' of *r*-process abundances observed from star to star.

1 Introduction

The potential to probe the origin(s) of the heaviest elements observed in nature has never been greater. Previously supernovae were thought to be responsible for the synthesis of these elements but most modern simulations suggest that core-collapse cases do not have the proper conditions to reach the heaviest observed nuclei (e.g. [1]). A currently accepted open window for supernovae to produce heavy species is a special case with high magnetic fields known as magneto-rotationally driven supernovae, however many simulations suggest this environment struggles to reach the actinide elements (e.g. [2, 3]). A long discussed possible site is that of neutron star merger events (e.g. [4]) since neutron-rich material can be dynamically lifted from the surface of the neutron star(s) forming tidal tails during their approach or expelled more violently at the collision interface. Additionally an accretion disk can later form around the merger remnant and eject reprocessed material. In such conditions the rapid neutron capture process, *r* process, has the possibility to synthesize the heaviest species known to occur in nature.

*e-mail: nvassh@nd.edu

The era of multi-messenger astronomy now permits unprecedented insights into astrophysical events such as mergers, as evidenced by LIGO/VIRGO's ability to direct the telescope community to perform detailed electromagnetic follow-up for merger events such as GW170817 (e.g. [5–7]). Although this event broke ground in providing direct evidence for the production of lanthanide elements, it remains unclear whether species heavier than this such as gold, platinum, and the actinides were produced. An interesting way to search for evidence that the heaviest species were produced is to look to identify signatures of fission in astrophysical observables since if the actinides have been reached then lighter species such as gold must have been synthesisized as well. However investigating fission in the *r* process currently faces challenges from a general lack of understanding for the properties of the neutron-rich actinides. In this proceedings, we summarize the state of the nuclear data used for fission in *r*-process calculations and discuss currently identified potential signatures of fission within astrophysical environments.

2 A brief discussion of fission rates for neutron-rich nuclei

There are a limited number of available datasets for the fission reactions of neutron-rich nuclei (see [8–12] and references therein). Since the *r* process operates in a regime far from current experimental data, simulation results must be reported as a function of the theoretical model applied. There are only a handful of fission barrier predictions for neutron-rich nuclei in the literature. All available models predict fission barriers to become low for many neutron-rich nuclei of importance to the *r* process (see Figure 14 of [8] for a comparison of four publicly available barrier height sets). We note that although different barrier treatments imply differing reaches for ultimately how heavy of species the *r* process can produce [8], applications of rates determined from currently available fission barriers suggest that the *r* process does not populate the predicted super heavy element island of stability around $Z = 114$ and $N = 184$ [13]. For determining the abundances of *r*-process nuclei, currently calculations suggest neutron-induced and β-delayed fission to be of the greatest importance, with neutron-induced fission most commonly being predicted to be the means by which the reach of heavy element synthesis is terminated. However [13] highlighted the importance of β-delayed fission during the late-times which set *r*-process abundances and for the first time predicted that multi-chance β-delayed fission can occur for neutron-rich nuclei.

The limited availability of calculations for the fission properties of neutron-rich nuclei means prescriptions are sometimes not applied with self-consistent barriers for all fission reaction and decay channels. Since this can lead to nucleosynthesis results which may show features which are in actuality an artifact of such an inconsistency it is important to push toward a self-consistent implementation of fission properties in the *r* process. To apply rates which have considered the same set of fission barriers in both the neutron-induced and β-delayed fission processes which dominate how the synthesis of heavy elements proceeds, we use the LANL suite of Hauser-Feshbach codes (see [8] and references therein). Using four distinct theoretical fission barrier sets, in [8] we examined exactly which nuclei most influence *r*-process predictions for the final abundances from an astrophysical process such as merger tidal-tail ejecta which can robustly produce fissioning species. We found many of these nuclei to be far from current experimental reach and therefore highlighted the need for progress in theoretical descriptions. However, since the fissioning nuclei of most importance are those populated as β-decay moves the general population of *r*-process nuclei toward stability [8, 14], rather than those in the most neutron-rich regions near the dripline, dedicated future experiments may be able to close in on some of these species.

With respect to spontaneous fission, the rates available are often simple parameterized formulas fit to a limited set of data. This means such descriptions can disagree by orders of

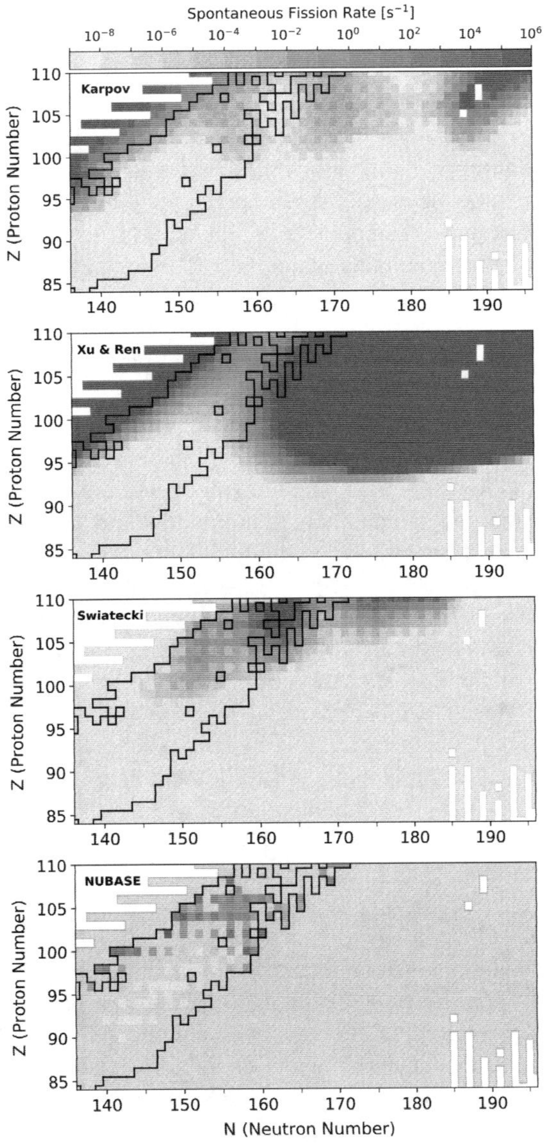

Figure 1. Spontaneous fission rates assuming FRLDM barriers [15] along with the phenomenological formula of Karpov *et al.* [16] (top panel) as compared to rates given by Xu & Ren [18] (second panel) and Swiatecki [17] (third panel) parameterizations. The bottom panel shows experimental rates given NUBASE2016 [19] half-lives and branching ratios for comparison. The black outline shows the extent of NUBASE half-life data and the grey region shows the FRDM2012 [20] neutron dripline.

magnitude when their prescription is extrapolated into neutron-rich regions (see Fig. 1). In our *r*-process calculations we apply spontaneous fission rates determined by the phenomenological equation of [16] which depends on fissility (Z^2/A) and barrier height. These spontaneous fission rates are very low until reaching $Z > 100$. The power law fit to fissility introduced by Swiatecki [17] (which also includes a shell correction term taken here to be the microscopic energy predicted by FRDM2012 [20]) also concentrates its high fission rates at $Z > 100$. In contrast, the phenomenological description of [18], which is parameterized as a function of fissility, $N - Z$, $N - 126$, and $Z - 82$, has dramatically different behavior, predicting a very large region of fast rates at $Z \gtrsim 94$ just above the $N = 184$ predicted shell closure. The bottom panel of Fig. 1 permits a comparison between these parameterized spontaneous fission fits and experimental data. Such parameterized descriptions clearly struggle to reproduce the rather abrupt order of magnitude variations seen in the experimental rates. While attempts to apply more physically meaningful approaches for the spontaneous fission rates of neutron-rich nuclei have been made in recent years [10], ultimately the availability of theoretical data for this process remains scarce. We look forward to theoretical calculations and experimental measurements to better inform the treatment of spontaneous fission in the *r* process in the future.

Since most descriptions show spontaneous fission to play its largest role at high proton number, consistent with other studies we find that *r*-process material tends to encounter spontaneous fission at late times when the abundances of almost all *r*-process nuclei are already finalized. However such late-time fission can have an impact on other *r*-process observables via the heat that a $\sim$ 200 MeV Q-value will transfer to the astrophysical environment. For instance, [21] showed that the spontaneous fission of the long-lived species ^{254}Cf has the potential to impact merger light curves. The light curve response to the spontaneous fission heating from ^{254}Cf takes place on the order of 10 to 100 days and is presently the only way possible to confirm that a merger event produced elements heavier than lanthanides such as gold and uranium. However whether or not ^{254}Cf can have such an impact depends on the fission barrier model assumed [8] since the nuclei on the β-decay path feeding this species can be depopulated from neutron-induced or β-delayed fission. Therefore, advancements in the theoretical description for the fission rates of neutron-rich isotopes is of great importance in understanding multi-messenger signals. At the present it is unclear whether other nuclei accessible to the *r* process with spontaneous fission half-lives on the order of days exist in neutron-rich regions.

3 Impact of fission yields on *r*-process abundances

In very neutron-rich conditions, fission fragment distributions play a crucial role in determining the abundances of lighter elements near the $A \sim 130$ second *r*-process peak (such as tellurium ($Z = 52$)). This is largely due to a significant abundance of species present in the fissioning regions after the fast neutron capture of lighter elements pushes out to very high mass numbers just before the neutron flux is exhausted. These high mass species then deposit lower mass number daughter products into the main *r*-process pattern (between the second and third ($A \sim 195$) *r*-process peaks) when they fission. In our calculations we consider fission yield daughter products at the 0.01% level and above. We have found that the order of magnitude threshold at which *r*-process abundances are sensitive to the fission yields of neutron-rich nuclei is at the 0.1% level since calculations which cut off yields at the 1% level produce different abundance results.

As can be seen from the comparison in Fig. 2, the prescription for fission fragment distributions can lead to differences as large as an order of magnitude in the *r*-process abundances. In addition to the impact from the primary fragment yield treatment, we considered the sensitivity of *r*-process abundances to the excitation energy dependence of the fission yields for each distinct fission channel (neutron-induced, β-delayed, and spontaneous) in [8] using the GEF2016 code [22]. Prior to this work, common practice for *r*-process calculations was to apply one yield set to all channels and neglect the excitation-energy dependence of fission yields. Additionally we considered the impact of prompt neutron emission via direct comparisons of results with GEF and FREYA [23, 24] for neutron-rich nuclei. As shown in Fig. 2, we found that considering yields which have a non-vanishing excitation energy was of greater impact on *r*-process abundances than the de-excitation treatment which determines prompt neutron emission.

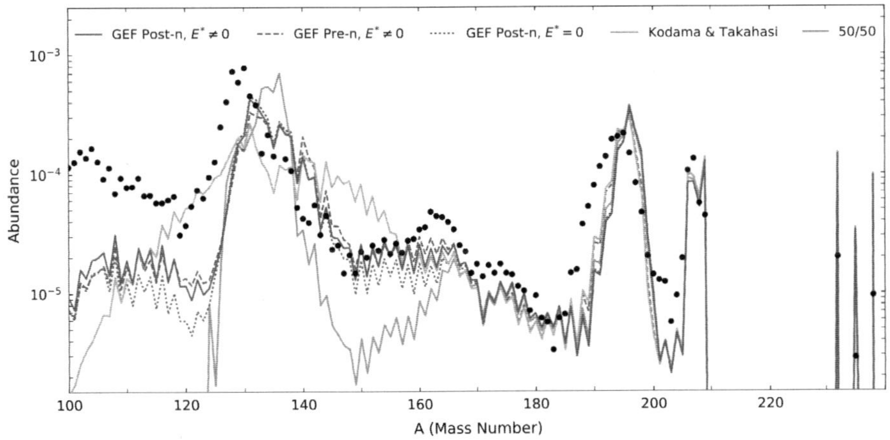

Figure 2. Nucleosynthesis calculation abundances at 1 Gyr as a function of mass number for cold dynamical ejecta conditions as can be found in a merger tidal-tail [25] using different prescriptions for the fission yields of neutron-rich nuclei: GEF2016 [22] (blue), Kodama & Takahashi [26] (yellow), and 50/50 symmetric yields where it is assumed that the nucleus simply splits in half (green). The dashed blue line examines the impact of ignoring prompt neutron emission by applying pre-neutron emission yields. The dotted blue line considers results when the excitation energy dependence of yields is ignored ($E^* = 0$). Black dots are the solar abundance data from [27].

4 Fission yields from macroscopic-microscopic theory in the *r* process

Very few theoretical approaches predicting fission yields have been applied to neutron-rich nuclei [11, 28, 29] and only a subset have considered all of the neutron-rich isotopes needed to perform a nucleosynthesis calculation. Therefore the majority of calculations in the literature make use of data-driven phenomenological models, such as the GEF treatment presented in the last section as well as the ABLA code (the precursor of GEF) and the Wahl systematics (see [8] for references and a literature overview of fission yield applications in the *r* process). Recently, fission yields calculated using the FRLDM macroscopic-microscopic approach [15] have been reported for the broad range of neutron-rich isotopes of relevance in the *r* process [28]. These fragment distributions have a wider range of daughter products than previously predicted by phenomenological extrapolations such as GEF2016, as can be seen in Figure 3.

The elemental *r*-process abundances using FRLDM fission yields, as compared to applying a simple 50/50 symmetric split for fissioning nuclei, are shown in Fig. 4 for several distinct astrophysical conditions. A merger tidal tail is often found by simulation to be very neutron-rich and therefore can easily synthesize up to the fissioning isotopes. Calculations for such a condition demonstrate the ability of the wide yields of FRLDM to not only populate the main *r*-process nuclei between the second and third peaks, but also significantly produce the light precious metals such as palladium ($Z = 46$) and silver ($Z = 47$). Such co-production of light precious metals and lanthanide elements ($Z = 57$ to $Z = 71$) is not observed for more simple fission yield prescriptions such as 50/50 splits. Also shown are

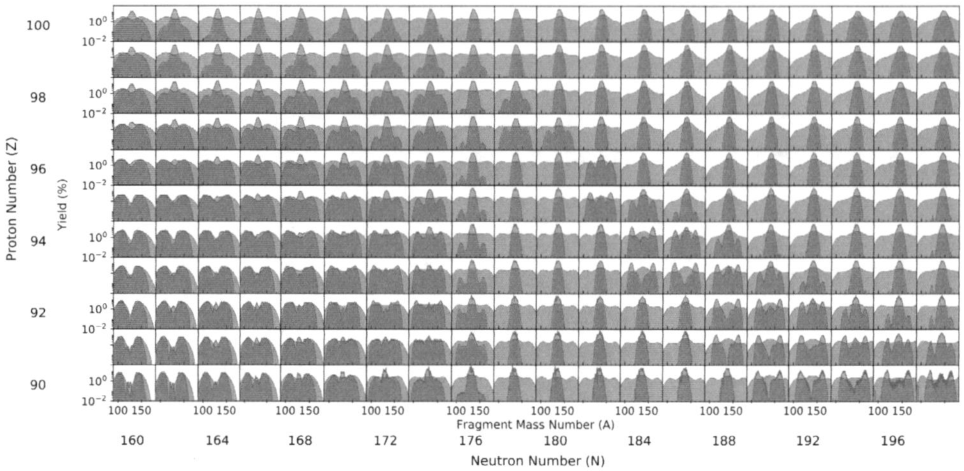

Figure 3. The fission yields predicted for some neutron-rich nuclei of relevance in the *r* process given recent results which make use of the macroscopic-microscopic theory of FRLDM [28] (red) as compared to the data-driven, phenomenological calculations of GEF2016 (neutron-induced fission with an incident neutron energy of 0.1 MeV) [22] (blue).

abundances given three distinct accretion disk wind trajectories, all of which can be found in the simulation of [30]. In this wind simulation, most trajectories produce abundances similar to the 'weak' *r*-process case which does not populate nuclei beyond the second *r*-process peak. It is clear from the case examining 50/50 yield results that without contributions to the light precious metals from fission deposition the only way to populate such species at a given astrophysical site is for some of the ejecta to undergo solely a weak *r* process. Given naturally occurring variations at astrophysical sites such as the mass of the neutron stars and the nature of the merger remnant, the amount of weak *r*-process ejecta relative to ejecta which produces heavier species such as the lanthanides and actinides can also naturally vary. However if conditions which access fissioning nuclei are significantly present at a site, the co-production of weak *r*-process elements and lanthanides gives final abundances which are more immune to variances in weak versus main *r*-process ejecta. This stabilizing effect on the final abundances could help explain the so-called 'universality' of elemental abundances observed in *r*-process enhanced, metal-poor stars [31].

5 Possible signature of co-production via fission in *r*-process enhanced, metal-poor stars

Observations of the elemental abundances in our Sun and other stars can provide insights as to the nature of the site(s) producing heavy elements. Examining metal-poor stars which are rich in *r*-process material (i.e. r-I and r-II stars [32]) reveals the elemental production from one to few nucleosynthesis events since such stars have not yet been significantly polluted by events such as supernovae which inject metal-rich material into the interstellar medium. Elemental patterns for such stars show a remarkable consistency with each other and the content of *r*-process material in our Sun from barium ($Z = 56$) through the third *r*-process peak [27]. This 'robustness' or 'universality' has been discussed as a potential signature of fission in the *r* process (see [31] and references therein) since fission deposition back into

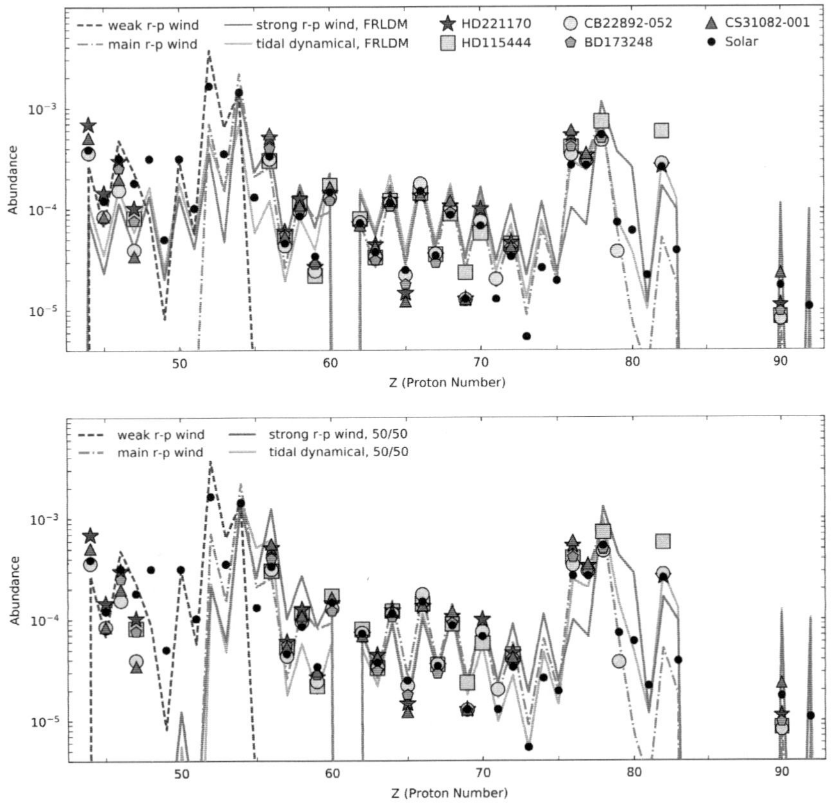

Figure 4. The elemental abundances predicted by nucleosynthesis calculations at 1 Gyr for dynamical merger tidal-tail ejecta [25] (orange) as compared to three distinct conditions all present within a simulation of accretion disk wind ejecta [30]: a weak r process which does not synthesize elements beyond the second r-process peak (purple dashed), a main r process which mostly populates nuclei between the second and third r-process peaks and weakly produces actinides (blue dashed-dotted) and a strong r process which robustly reaches very neutron-rich, fissioning species. The top panel shows results using FRLDM fission yields and the bottom panel shows these cases using 50/50 symmetric splits. The solar and stellar abundance data is from [27].

lighter elements near $A \sim 130$ can potentially wash away differences in the initial conditions prior to the synthesis of fissioning species.

In considering a possible connection between fission and universality, it is important to identify what feature of stellar abundances universality statements are in fact referencing. For instance, if universality refers to similarities in the r-I and r-II stellar data for the abundances of r-process nuclei, fission may not be the process producing this since in [8] we demonstrated that astrophysical conditions which robustly reach fissioning nuclei can have varying abundance patterns. This is in part due to distinct conditions populating different fissioning species (implying distinct fission yield distributions set abundances when depositing daughter products). Differences in conditions also determine how much of the second peak is formed by late-time fission as well as how much late-time neutron capture can shift the second peak. It is unclear how large such differences in abundances can be in an actual astrophysical environment hosting fission given the variance seen in the conditions reported by astrophysical simulations as well as the uncertainties in the fission barriers and fission yields of neutron-rich nuclei. Additionally it is unclear whether the uncertainties in stellar observations of elemental abundances overshadow the abundance differences referred to here from varying astrophysical conditions.

If instead universality statements are in reference to the abundances of lanthanides as compared to abundances of elements in the third *r*-process peak, fission can still be a means to explain the consistency in the stellar data since all fissioning conditions robustly produce the full range of *r*-process nuclei. However, some universality arguments solely consider lanthanide abundances relative to each other since elemental patterns here are indeed remarkably similar from star to star. In fact, some authors have made use of stellar data to determine the solar *r*-process element content for lanthanides such as hafnium ($Z = 72$) and gadolinium ($Z = 64$) [27]. Here we examine whether universality in the lanthanides alone can be linked to fission. We also consider whether comparisons between the light precious metals to the left of the second *r*-process peak and the lanthanides can shed light on the prospect of late-time fission deposition in the *r* process.

We show the stellar data [32] for the elemental ratios for some light heavy elements (Mo ($Z = 42$), Ru ($Z = 44$), Rh ($Z = 45$), Pd ($Z = 46$), and Ag ($Z = 47$)) as compared to the lanthanide element, Eu ($Z = 63$) in Fig. 5. We compare to the ratio predicted by nucleosynthesis calculations. Note in these figures we use log eps(Eu) (the europium abundance) to quantify the *r*-process enrichment and the stellar values are that reported by observation. The values shown for the nucleosynthesis calculations must be scaled according to the absolute amount of europium, which depends on the amount of ejected matter from the astrophysical scenario considered, and this is presently unknown. Therefore in considering the agreement between stellar data and calculated values, only the ratios on the y-axis can be compared in a physically meaningful way.

Left panels of Figure 5 use the wind calculations of [30] as an example of conditions which primarily produce a weak *r* process and most significantly populate nuclei lighter than those in the second *r*-process peak. Also shown are results with dynamical ejecta from [25] for which all conditions robustly reach fissioning nuclei. Right panels consider the nucleosynthesis productions given a more modern dynamical ejecta simulation from [33] which reports a wider range of conditions thereby producing both a weak and full *r* process. Right panel nucleosynthesis results are color-coded by the total integrated fission flow (rate times abundance) in order to highlight the conditions present which most strongly access fissioning nuclei. The spread in the observational data for elements such as Mo and Ru suggests these elements to been produced independently of Eu, possibly by sites which do not synthesize past the second *r*-process peak. For the heaviest of the light heavy elements shown here, Pd and Ag, the flat trend seen in the stellar data suggests that such elements are co-produced with Eu in the astrophysical events which enriched these stars. Our nucleosynthesis calculations see conditions which are most consistent with observed ratios to be those which co-produce light precious metals and lanthanides via late-time fission deposition.

In Fig. 6 we examine elemental ratios of some lanthanide elements (La ($Z = 57$), Pr ($Z = 59$), Nd ($Z = 60$), Dy ($Z = 66$), and Yb ($Z = 70$)) relative to another lanthanide element Eu. The flat trends in the stellar data of r-I and r-II stars show the similarities in the observational data for these ratios and demonstrate the type of trend which is indicative of co-production (as was suggested by the Ag and Pd trends in Fig. 5). Here we see it is not only conditions which reach fissioning nuclei which can accommodate these trends since conditions which instead produce a main *r* process but do not reach fissioning species also often predict a ratio comparable to the data. Therefore if universality is to be used as a probe of whether fission has played a significant role in setting the final abundances, the elemental ratios of lanthanides relative to each other are not the most relevant feature since fission is not required to accommodate the stellar trends. Rather it is the relative ratios of lanthanides to other regions of the abundance patterns, such as the third peak or light precious metals, such as Ag and Pd, which could reveal the participation of fission in the astrophysical scenario which produced these nuclei.

6 Conclusions

Fission is a unique probe of the astrophysical origins of the heaviest species observed in nature. Since the participation of this exotic and energetic process implies that elements such as gold have been synthesized prior to the building of the actinides, fission signatures are a means to understand whether an astrophysical event has produced all *r*-process nuclei. Here we discussed the challenges in quantifying the impact of fission in the *r* process due to the uncertainties in the fission rates and fission fragment distributions of neutron-rich nuclei. We discussed potential signatures of fission which were able to be identified when applying currently available models, such as the heating impact from late-time spontaneous fission on merger light curves. This highlights the need for both experiment and theory to extend their current reach into neutron-rich regions. We showed that recent calculations for fission yields using macroscopic-microscopic theory predict distributions which are distinct from past models via their enhanced width and therefore range of deposition in the *r* process. We discussed the possible connection between such deposition into the light precious metals and the so-called 'robustness' or 'universality' of the abundances observed for metal poor stars rich in *r*-process elements. We pointed out that attempts to link fission and universality should be subject to more quantified statements. For instance, the universality observed amongst lanthanide element abundances relative to each other does not require a participation from fission. However universality discussions should be extended to consider the ability of fission to co-produce the light precious metals, such as Ag and Pd, along with the lanthanide elements since fissioning conditions most closely and robustly reproduce [Ag/Eu] and [Pd/Eu] stellar trends. Thus although there remain many mysteries as to the astrophysical origins of heavy elements, including whether or not fission significantly occurs in *r*-process environments, luckily due to the extreme impacts of fission on *r*-process observables, such questions become more approachable in this new era of multi-messenger events and precision observations.

Acknowledgements

The work of N.V., M.R.M., R.S., and R.V. was partly supported by the Fission In R-process Elements (FIRE) topical collaboration in nuclear theory, funded by the U.S. Department of Energy. Additional support was provided by the U.S. Department of Energy through contract numbers DE-FG02-95-ER40934 (R.S. and T.M.S.), and DE-SC0018232 (SciDAC TEAMS collaboration, R.S. and T.M.S). T.M.S. was supported in part by the Los Alamos National Laboratory Center for Space and Earth Science, which is funded by its Laboratory Directed Research and Development program under project number 20180475DR. R.S. also acknowledges support by the National Science Foundation Hub (N3AS) Grant No. PHY-1630782. The work of R.V. was performed under the auspices of the U.S. Department of Energy by Lawrence Livermore National Laboratory under Contract DE-AC52-07NA27344. M.R.M. was supported by Los Alamos National Laboratory operated by Triad National Security, LLC, for the National Nuclear Security Administration of U.S. Department of Energy (Contract No. 89233218CNA000001). This work was partially enabled by the National Science Foundation under Grant No. PHY-1430152 (JINA Center for the Evolution of the Elements).

References

[1] J. Bliss, A. Arcones, F. Montes, *et al.*, Journal of Physics G Nuclear Physics **44**, 054003 (2017)

[2] C. Winteler, R. Käppeli, A. Perego, *et al.*, ApJL **750**, L22 (2012)

[3] P. Mösta, L. F. Roberts, G. Halevi, *et al.*, ApJ **864**, 171 (2018); G. Halevi & P. Mösta, MNRAS **477**, 2366 (2018)

[4] J. M. Lattimer, F. Mackie, D. G. Ravenhall, *et al.*, ApJ **213**, 225 (1977)

[5] P. S. Cowperthwaite, E. Berger, V. A. Villar, *et al.*, ApJL **848**, L17 (2017)

[6] B. P. Abbott, R. Abbott, T. D. Abbott, *et al.* (LIGO Scientific Collaboration and Virgo Collaboration), Phys. Rev. Lett. **119**, 161101 (2017)

[7] D. Kasen, B. Metzger, J. Barnes, *et al.*, Nature **551**, 80 (2017)

[8] N. Vassh, R. Vogt, R. Surman, *et al.*, Journal of Physics G Nuclear Physics **46**, 065202 (2019)

[9] M. Eichler, A. Arcones, A. Kelic, *et al.*, ApJ **808**, 30 (2015)

[10] S. A. Giuliani, G. Martínez-Pinedo & L. M. Robledo, Phys. Rev. C **97**, 034323 (2018)

[11] S. Goriely, Eur. Phys. J. A **51**, 22 (2015)

[12] S. Shibagaki, T. Kajino, G. J. Mathews, *et al.*, ApJ **816**, 79 (2016)

[13] M. R. Mumpower, T. Kawano, T. M. Sprouse, *et al.*, ApJ **869**,14 (2018)

[14] T. M. Sprouse, *et al.*, this proceedings

[15] P. Möller, A. J. Sierk, T. Ichikawa, *et al.*, Phys. Rev. C **91**, 2, 024310 (2015)

[16] A. V. Karpov, V. I. Zagrebaev, Y. Martinez Palenzuela, *et al.*, Int. J. Mod. Phys. E **21**, 1250013-1-1250013-20 (2012)

[17] W. J. Swiatecki, Physical Review **100**, 937 (1955)

[18] C. Xu & Z. Ren, Phys. Rev. C **71**, 014309 (2005)

[19] G. Audi, F. Kondev, M. Wang, *et al.*, Chin. Phys. C **41**, 030001 (2017)

[20] P. Möller, A. J. Sierk, T. Ichikawa, *et al.*, At. Data Nucl. Data Tables **109**, 1 (2016)

[21] Y. Zhu, R. T. Wollaeger, N. Vassh, *et al.*, ApJL **863**, L23 (2018)

[22] K.-H. Schmidt, B. Jurado, C. Amouroux, *et al.*, Nucl. Data Sheets **131**, 107 (2016)

[23] J. M. Verbeke, J. Randrup, & R. Vogt, Comp. Phys. Comm. **191**, 178 (2015); J. M. Verbeke, J. Randrup, & R. Vogt, Comp. Phys. Comm. **222**, 263 (2018)

[24] R. Vogt, *et al.*, this proceedings

[25] S. Rosswog, T. Piran, & E. Nakar, MNRAS **430**, 2585 (2013)

[26] T. Kodama & K. Takahashi, Nucl. Phys. A **239**, 489 (1975)

[27] C. Sneden, J. J. Cowan, & R. Gallino, ARA&A **46**, 241-288 (2008)

[28] M. Mumpower, P. Jaffke, M. Verriere, *et al.*, arXiv e-prints, arXiv:1911.06344 (2019)

[29] J. Sadhukhan, S. Giuliani, Z. Matheson, *et al.*, arXiv e-prints, arXiv:2001.08616 (2020)

[30] O. Just, A. Bauswein, R. A. Pulpillo, *et al.*, MNRAS **448**, 541 (2015)

[31] N. Vassh, M. R. Mumpower, G. C. McLaughlin, *et al.*, arXiv e-prints, arXiv:1911.07766 (2019)

[32] A. Abohalima, & A. Frebel, ApJS **238**, 36 (2018)

[33] D. Radice, A. Perego, K. Hotokezaka, *et al.*, ApJ **869**, 130 (2018)

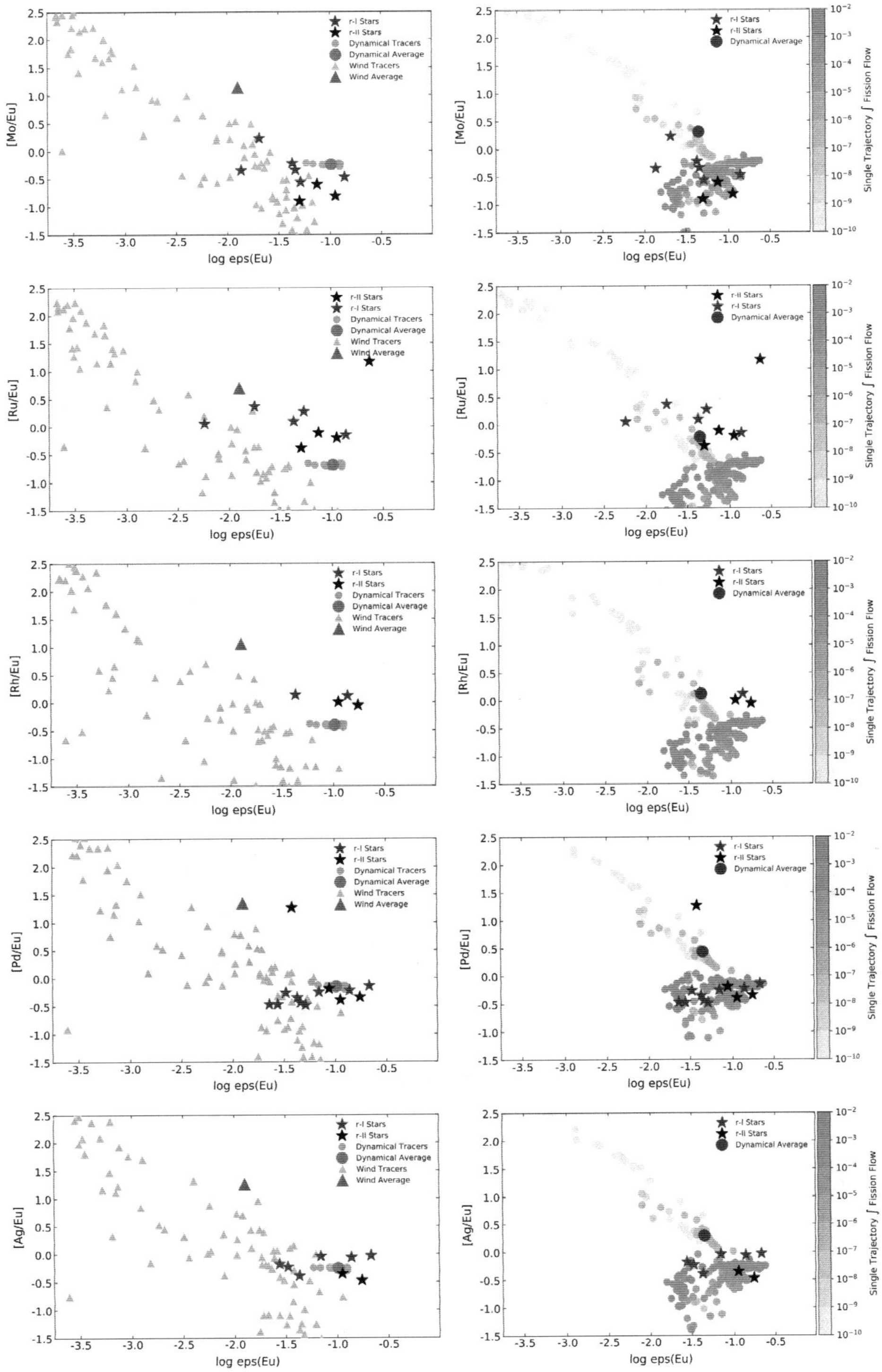

Figure 5. Calculation results for the elemental ratios of light heavy elements (Mo, Ru, Rh, Pd, and Ag) found to the left of the second *r*-process peak relative to the lanthanide element Eu found to the right of the peak (a subset of such plots was shown in [31]). Observational values for stars which are metal-poor, yet rich in *r*-process elements are shown for comparison. See text for details.

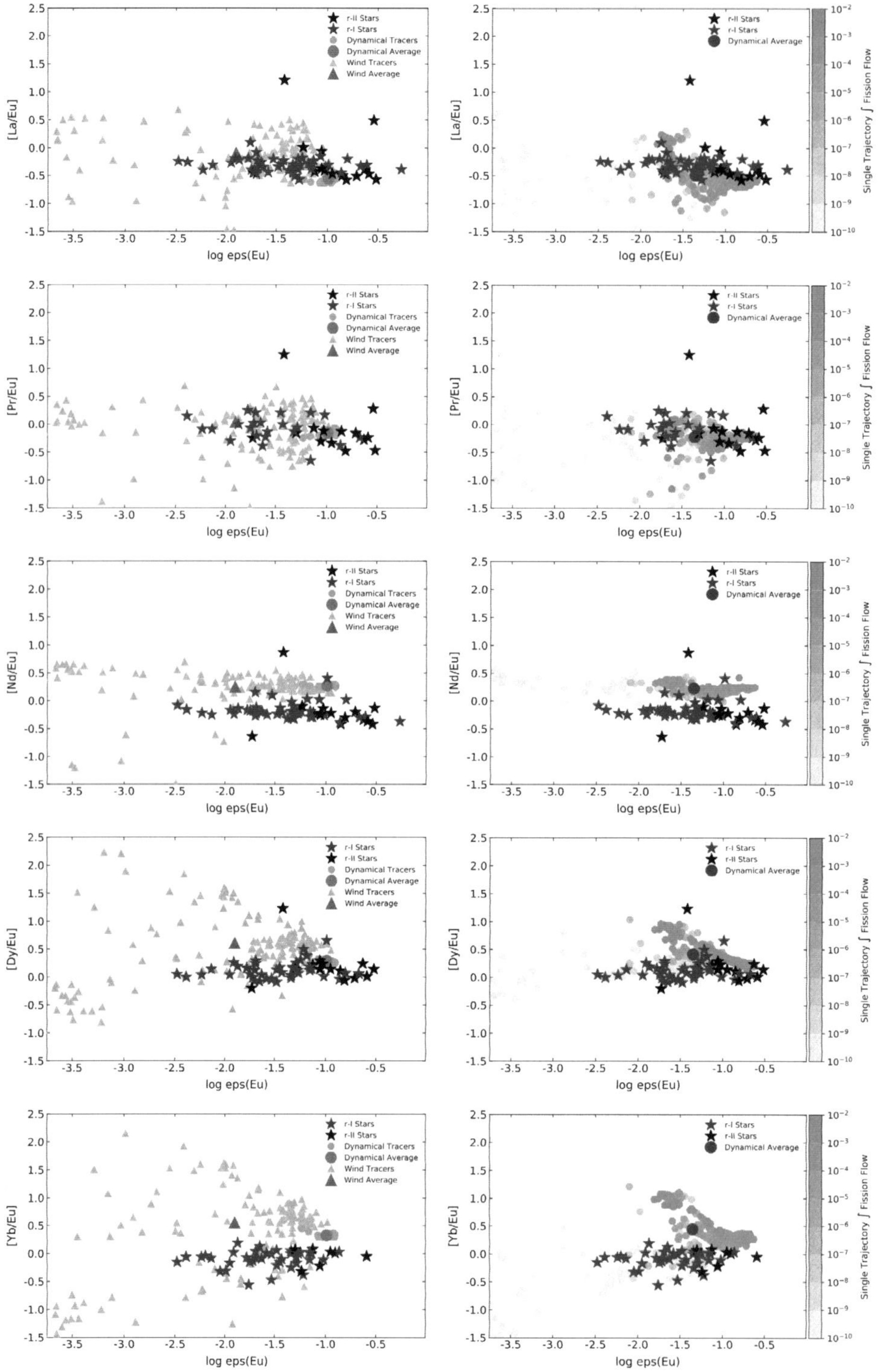

Figure 6. Same as Fig. 5 but for lanthanide species (La, Pr, Nd, Dy, and Yb) relative to another lanthanide element, Eu (a subset of such plots was shown in [31]).

From fission yield measurements to evaluation: new statistical methodology applied to ^{235}U(n_{th}, f) mass yields

Grégoire Kessedjian[1,*], *Sidi Mohamed* Cheikh[1], *Abdelazize* Chebbouboi[2] and *Olivier* Serot[2]

[1]LPSC, Université Grenoble-Alpes, CNRS/IN2P3, F-38026 Grenoble Cedex, France
[2]CEA, DEN, DER, SPRC, LEPh, Cadarache center, F-13108 Saint Paul lez Durance, France

Abstract. The study of fission yields has a major impact on the characterization and understanding of the fission process and is mandatory for reactor applications. The mass and isotopic yields of the fission fragments have a direct influence on the predictions of fuel burn-up and decay heat. Moreover, these data are requested for other studies as delayed neutron evaluation, antineutrino flux assessment or reactor program. Today, the lack of covariance matrix associated to evaluated fission yields induces overestimated uncertainties of mass yields since these observables result from the sum of isotopic and isomeric yields. Our collaboration starts a new program in the field of the evaluation of fission products in addition to the current experimental program. The goal is to define a new methodology of evaluation based on statistical tests in order to provide the best estimation with consistent sets of measurements. A ranking of solutions with associated covariance based on Shannon's entropy criterion is proposed for the mass yields from ^{235}U(n_{th}, f) reaction.

1 Introduction

Fission yields evaluation represents the synthesis of experimental and theoretical knowledges in order to perform the best estimation of mass, isotopic and isomeric yields. Nevertheless the estimation of thes observables are drastically based on experimental data since the modelling of fission process is not predictive. Today, the output of fission yields evaluation is available as a function of isotopic and isomeric yields. As a consequence, mass yields are the sum of isobar nuclei and their quadratic sum to deduce uncertainties. So without any correct covariance matrix, mass yields uncertainties are greater than isotopic yields. This consequence is in contradiction with experimental knowledges where the abundance of mass yields measurements is clearly dominant. Thus, we expect the uncertainties on this latter observable to be lower than those of isotopic yields. Covariance matrix assessment depends on the evaluation process and its validity assumes that all measurements are statistically in agreement. These last years, different covariance matrices have been suggested but the experimental part of those are neglected in covariance evaluation [1][2][3][4] or applications [5][6]. In the first part we present an assessment methodology based on statistical test. The consideration of experimental data is crucial in the definition of the covariance of evaluations where models are not predictive. A large range of data are listed in the EXFOR data bank but a lot of them cover partial mass range. Data are also provided for different incident neutron

*e-mail: kessedjian@lpsc.in2p3.fr

Table 1. Numbers of common masses measured $n_{\hat{A}}$ for the 8 datasets selected.

Datasets	Cumulative measurement number	Cumulative mass number	Index	0	1	2	3	4	5	6	7
Maeck78	37	37	0	37							
Thierens75	74	48	1	26	37						
Diiorio77	138	64	2	37	36	64					
Bail07	168	67	3	17	19	28	30				
Tsoukatos68	232	67	4	37	36	64	28	64			
Tsoukatos68b	260	67	5	23	20	28	13	28	28		
Rosman83	267	72	6	0	0	1	2	1	0	7	
Mathiews83	279	77	7	0	0	1	2	1	0	7	12

energy with different mass resolutions. The merge of data could generate the non-unique solutions according to consistent datasets and will be presented in the second part. In the last part, a ranking of solutions with associated covariance is proposed according to the Shannon's entropy criterion.

2 Experimental datasets

For this first work, the development of the evaluation method is focused on $^{235}U(n_{th}, f)$ reaction because of a large number available measurements. Thus, for this reaction the EXFOR [7] database permits to cover all the mass range of fission products. Absolute mass yields Y(A) is obtained using the self-normalization of this observable according to the equation:

$$\sum_A Y(A) = \Omega \tag{1}$$

where Ω is the normalization factor, here $\Omega = 2$. Unfortunately, for most of these measurements, only statistical uncertainties is provided and systematic uncertainties are estimated in the best cases. Thus the estimation of covariance matrix is non-obvious and represents an important work which will be described in a future work.

2.1 Statistical test on the compatibility of available data

In order to reduce the data and to merge all the measurements for a given mass, it is necessary to test the compatibility of the data. Two kind of data are presented: i) the full range mass yield measurements but not necessarily with sufficient mass resolution ($3\sigma_A < 1amu$), ii) uncomplete mass range generating relative measurements or relative normalizations by the authors. Through the EXFOR [7] database, we chose to test the methodology only on 8 datasets: W.J. Maeck et al. [8], G. Diiorio et al. [9], H. Thierens et al. [10], A. Bail et al. [11], M.P. Tsoukatos [12] with two different datasets, K.J. Rosman [13], C.K. Mathiews [14]. These data correspond to 279 measurements over 77 masses (from A=77 to A=154 with a lack of A=122 in the data used in this work). With this selection, we cover both peaks, allowing the absolute normalization of our evaluation. Thus, assuming independent Gaussian distributions associated to the measurements without explicit information on correlation data, we can calculate the χ^2 using the n_A common measured mass number. This value is compared to the limited χ^2 value (χ^2_{lim}) given for a 99.5% confidence level. In practice, we calculate the

Table 2. Relative standard deviations of normalization factor k_{ij} between all datasets.

$\sigma(k_{ij})$ (%)	0	1	2	3	4	5	6	7
0	0.0	1.4	0.2	0.7	0.2	0.3	-	-
1	1.4	0.0	1.2	1.8	1.2	1.5	-	-
2	0.2	1.2	0.0	0.6	0.2	0.2	12.2	33.8
3	0.7	1.8	0.6	0.0	0.6	0.8	3.8	28.2
4	0.2	1.2	0.2	0.6	0.0	0.2	12.2	33.8
5	0.3	1.5	0.2	0.8	0.2	0.0	-	-
6	-	-	12.2	3.8	12.2	-	0.0	25.0
7	-	-	33.8	28.2	33.8	-	24.9	0.0

$P-values$ corresponding to the integral on $[\chi^2;\infty]$ range of the χ^2 distribution for $(n_{\hat{A}}-1)$ degrees of freedom. The usable data are only those which pass the χ^2 test and common mass set $\hat{A}$ are defined as following:

$$\{\hat{A}\} = \{\{A\}_i \cap \{A\}_j \setminus P-values\ (\chi^2_{ij}) > 1 - CL\} \tag{2}$$

where $\{A\}_i$ (respectively $\{A\}_j$) are the measured masses of the ith (respectively j^{th}) dataset and CL is the confidence level chosen at CL=0.995 for this work. The common mass number $n_{\hat{A}} = Card(\{\hat{A}\})$ is presented in table 1. Formally, the comparison of each dataset $\mathbf{N}_j(A)$ to the reference one $\mathbf{N}_i(A)$ generates $P-values$ lower than $1 - CL = 0.005$ for all data set. So we introduce a cross-normalization factor $k_{i,j}$ to maximize the number of measurements in agreement considering all measurements as relative ones. This comparison is described by the $\mathbf{C_{ij}}$ vector following:

$$\mathbf{C_{ij}} = \mathbf{k_{ij}}.\mathbf{N_j} - \mathbf{N_i} \tag{3}$$

The normalization factors is obtained considering the minimum of the generalized χ^2_g over the $\{\hat{A}\}$ measurements in agreement:

$$\chi^2_g(k_{ij}) = \mathbf{C^T_{ij}}.\mathbf{Cov^{-1}}.\mathbf{C_{ij}} \tag{4}$$

where $\mathbf{Cov^{-1}}$ is the inverse covariance matrix associated to $\mathbf{C_{ij}}$. nevertheless, at this step without the covariance of the measurements, we consider that:

$$Cov(C_{ij}(A); C_{ij}(A')) = \left(Var(N_i(A) + k^2_{ij}.Var(N_j(A')\right).\delta_{AA'} \quad \forall A, A' \in \{\hat{A}\} \tag{5}$$

The standard deviations of the cross-normalization factors $k_{i,j}$ are presented in table 2. A discussion about usable data management is described in reference [15] [16].

2.2 Cross-correlations of usable data

The relative normalization of each jth dataset, $N_j(A)$, to the (i^{th}) reference one, $N_i(A)$, is defined by as follows:

$$\mathbf{R^i_j} = \mathbf{k_{i,j}}.\mathbf{N_j} \tag{6}$$

According to the perturbation theory [17], the covariance of two normalized measurements $R^i_l(A)$ and $R^i_j(A')$ is developed in the appendix (see Sect. Apppendix). For this study, without explicit experimental covariance matrices, most of components of variance-covariance are considering null:

$$Cov(N_j(A); N_l(A')) = Var(N_j(A)).\delta_{AA'}.\delta_{jl} \quad \forall j, l$$

For n measurements of mass A [18], the mean normalized mass rate $\bar{R}(A)$ is equal to:

$$\bar{R}(A) = W_1.R_1^i(A) + ... + W_n.R_n^i(A) = \left(\sum_{j=1}^{n} W_j.R_j^i(A)\right) \tag{7}$$

For n measurements of mass A and m measurements of mass A' [18], the covariance of mean normalized mass rates is equal to (see Sect. Apppendix, eq. 23):

$$Cov(\bar{R}(A); \bar{R}(A')) = \sum_{l,j=1}^{n,m} W_l \, C_{lj} \, W_j \tag{8}$$

with here $C_{lj} = Cov(R_l^i(A); R_j^i(A'))$ for two different masses A and A', thus $(n \times m)$ terms. The weights are defined (see Sect. Apppendix, eq. 24):

$$W_{j=1,n} = \left(\sum_{l}^{n} C_{lj}^{-1}\right) / \left(\sum_{l,j}^{n,n} C_{lj}^{-1}\right) \quad and \quad W_{l=1,m} = \sum_{j}^{m} C_{lj}^{-1} / \sum_{l,j}^{m,m} C_{lj}^{-1} \tag{9}$$

with here $C_{lj} = Cov(R_l^i(A); R_j^i(A))$ for a same mass A with $(n \times n)$ covariance terms (or for a same mass A' with $(m \times m)$ covariance terms). Fig. 1 presents the cross-correlation of $R_l^i(A)$ data for two different reference sets. We note that the intensity of the correlation depends drastically of the choice of normalization and the uncertainty of the normalization factor $\sigma(k_{ij})$ (see table 2).

3 Ranking of solutions based on Shannon's entropy

According to the normalization of mass yields (see eq.1), the generalized perturbation theory [17] allows to describe the variance-covariance matrix associated to the evaluation of the mass yields:

$$\frac{Cov(Y(A); Y(A'))}{Y(A).Y(A')} = \sum_{A''} S_{Y(A)\bar{R}(A'')} S_{Y(A')\bar{R}(A'')} \frac{Var(\bar{R}(A''))}{\bar{R}^2(A'')} + \sum_{A'',A'''} S_{Y(A)\bar{R}(A'')}.S_{Y(A')\bar{R}(A''')}.\frac{Cov(\bar{R}(A''); \bar{R}(A'''))}{\bar{R}(A'').\bar{R}(A''')} \tag{10}$$

with the sensitivity of mass yield Y(A) to mean mass rates $\bar{R}(A)$ and $\bar{R}(A')$:

$$S_{Y(A)\bar{R}(A)} = 1 - \frac{Y(A)}{\Omega} \quad and \quad S_{Y(A)\bar{R}(A')} = -\frac{Y(A')}{\Omega} \tag{11}$$

where Ω is the normalization factor, here $\Omega = 2$ (see eq. 1). Fig. 2 shows the results for two different reference sets and we remark that the structures of these correlations are strongly different. Thus, the result of the mass yields evaluation depends on the initial datasets but also the path of analysis. In order to discriminate all possible evaluations, the Shannon's entropy S_{Sh} is chosen as a useful criterion in order to assess the brewing of information [19]. It is given by the relation:

$$S_{Sh} = -\frac{1}{ln(2)} \sum_{k=1}^{n} P_i ln(P_i) \tag{12}$$

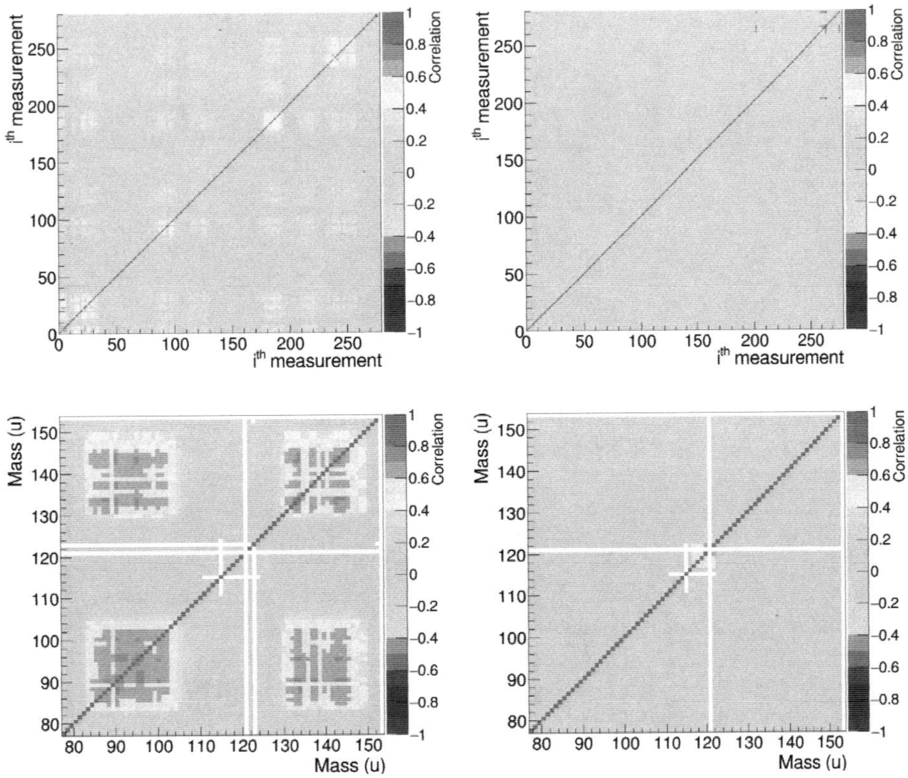

Figure 1. Preliminary results. (up, left) Cross-correlations for 8 datasets after normalization to the 3th set; (up, right) the same data in reference to the 4th set. These differences come from the uncertainties of the normalization factor $\sigma(k_{ij})$ which are not the same in both cases (see table 2). (down) Correlation of mean mass rate $Cov(\bar{R}(A); \bar{R}(A'))$ in reference to set 3 (left) and the set 4 (right).

Where n is the number of eigenvalues. We approximate the probability with the weight of each component of the eigenvalue decomposition to build a relative criterion. The weight of the information is provided according to the following equation:

$$P_i = \frac{EV_i}{tr(Corr)} \tag{13}$$

where $tr(Corr) = 77$ is the correlation matrix trace (in this study, 77 mass yields are evaluated).

4 Results and discussion

Results on pure experimental mass yields evaluation (modeless) are presented on Fig. 3. From all solutions, we note that the maximum of Shannon's entropy corresponds to the minimum of variances and correlations values. This results is consistent to the Cramer-Rao theorem which fixes the limits on minimal variances as the maximum of the Fischer's information. Shannon's entropy corresponds to another quantification of information of the analysis and we expect that the best searched solution corresponds to the minimum of variance-covariance and then

https://doi.org/10.1051/epjconf/202024205001

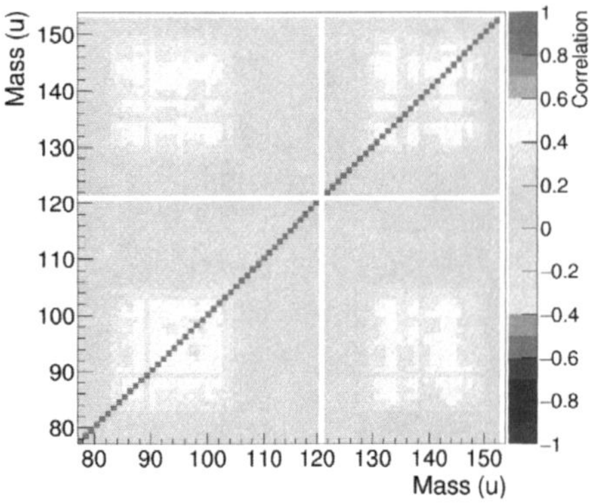

Figure 2. Preliminary results. (left) Mass yields correlations for 8 datasets after normalization to the 3th set; (right) the same in reference to the 4th set.

the maximum of information. In this work, experimental data consideration is crucial for the definition of the mass yields evaluation, the uncertainties and the correlations. The lack of experimental covariance could induce a lower estimation of evaluated mass yields uncertainties since the dealt information is overestimated. Correlations in the data limit the knowledge provided by a dataset. Then the perspective of this work is to build *a priori* experimental correlation matrix to fill the lacks in this analysis.

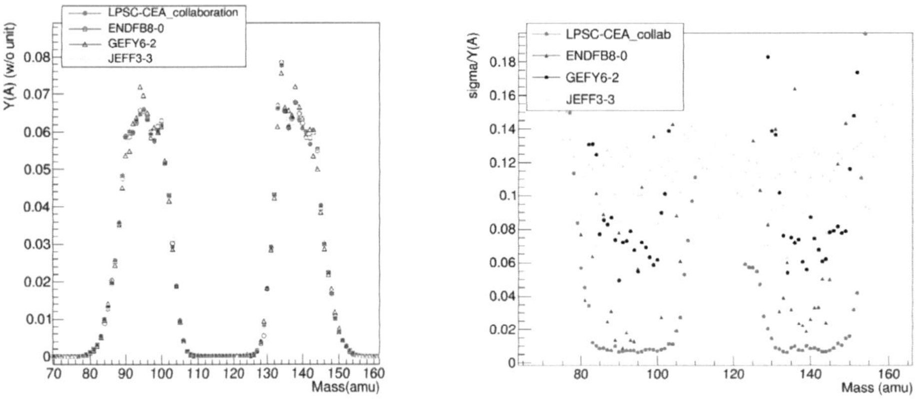

Figure 3. Preliminary results. (left) Mass yields evaluation proposed according to the maximum of Shannon's entropy (4th reference set) in comparison to JEFF-3.3 [20] and ENDF/VIII.0 [21] libraries and the calculation from GEFY6-2 [3]. (right) Relative standard deviation of evaluated mass yields in comparison to those of evaluations or GEFY6-2.

Acknowledgment

This work was supported by IN2P3, by the University of Grenoble Alpes and by "le défi NEEDS" and the (WP4) SANDA european project.

Appendix

The relative normalization of each jth dataset, $N_j(A)$, to the (i^{th}) reference one, $N_i(A)$, is defined as follows:

$$\mathbf{R^i_j} = \mathbf{k_{i,j}.N_j} \tag{14}$$

Based on the minimum of χ^2, the variance of normalization factor k_{ij} is given by:

$$\begin{aligned}
Var(k_{ij}) = & \frac{1}{K^2_{denom/j}} \sum_{\hat{A}} \left(\alpha^i_{ij}(\widehat{A})\right)^2 Var(N_i(\widehat{A})) + \frac{1}{K^4_{denom/j}} \sum_{\hat{A}} \left(\alpha^j_{ij}(\widehat{A})\right)^2 Var(N_j(\widehat{A})) \\
& + \frac{2}{K^2_{denom/j}} \sum_{\hat{A}} \sum_{\widehat{A'}>\widehat{A}} \left(\alpha^i_{ij}(\widehat{A})\right)\left(\alpha^i_{ij}(\widehat{A'})\right) Cov(N_i(\widehat{A}), N_i(\widehat{A})) \\
& + \frac{2}{K^4_{denom/j}} \sum_{\widehat{A}} \sum_{\widehat{A'}>\widehat{A}} \left(\alpha^j_{ij}(\widehat{A})\right)\left(\alpha^j_{ij}(\widehat{A'})\right) Cov(N_j(\widehat{A}), N_j(\widehat{A'}))
\end{aligned} \tag{15}$$

According to the perturbation theory, the covariance of two normalized measurements $R^i_l(A)$ and $R^i_j(A')$ is described by the following equation [17]:

$$\begin{aligned}
Cov(R^i_l(A), R^i_j(A')) = & \frac{N_l(A)N_j(A')}{K_{dénom/j}K_{dénom/l}} \sum_{\widehat{A}} \sum_{\widehat{A'}} \alpha^i_{ij}(\widehat{A})\alpha^i_{il}(\widehat{A'})Cov(N_i(\widehat{A}), N_i(\widehat{A'})) \\
& + \frac{N_l(A)N_j(A')}{K_{dénom/j}K^2_{dénom/l}} \sum_{\widehat{A}} \sum_{\widehat{A'}} \alpha^i_{ij}(\widehat{A})\alpha^l_{il}(\widehat{A'})Cov(N_i(\widehat{A}), N_l(\widehat{A'})) \\
& + \frac{N_l(A)N_j(A')}{K^2_{dénom/j}K_{dénom/l}} \sum_{\widehat{A}} \sum_{\widehat{A'}} \alpha^i_{il}(\widehat{A'}\alpha^j_{ij}(\widehat{A})Cov(N_j(\widehat{A}), N_i(\widehat{A'})) \\
& + \frac{N_l(A)N_j(A')}{K^2_{denom/l}K^2_{denom/j}} \sum_{\widehat{A}} \sum_{\widehat{A'}} \alpha^j_{ij}(\widehat{A})\alpha^l_{il}(\widehat{A'})Cov(N_j(\widehat{A}), N_l(\widehat{A'})) \\
& + \frac{k_{il}N_j(A)}{K_{dénom/j}} \left(\sum_{\widehat{A'}} \alpha^i_{ij}(\widehat{A'})Cov(N_l(A), N_i(\widehat{A'})) + \frac{1}{K_{dénom/j}} \sum_{\widehat{A'}} \alpha^j_{ij}(\widehat{A'})Cov(N_l(A), N_j(\widehat{A'})) \right) \\
& + \frac{k_{ij}N_l(A)}{K_{dénom/l}} \left(\sum_{\widehat{A}} \alpha^i_{il}(\widehat{A})Cov(N_j(A'), N_i(\widehat{A})) + \frac{1}{K_{dénom/l}} \sum_{\hat{A}} \alpha^j_{il}(\widehat{A})Cov(N_j(A'), N_l(\widehat{A})) \right) \\
& + k_{il}k_{ij}Cov(N_l(A), N_j(A'))
\end{aligned} \tag{16}$$

with:

$$\alpha^i_{ij} = \frac{N_j(\hat{A})}{\sigma^2_{ij}} \qquad \alpha^j_{ij} = \frac{(K_{denom/j}N_i(\hat{A}) - 2K_{num/j}N_j(\hat{A}))}{\sigma^2_{ij}} \tag{17}$$

$$K_{num/j} = \sum_{\hat{A}} \frac{N_i(\widehat{A})N_j(\widehat{A})}{\sigma^2_{ij}} \qquad K_{denom/j} = \sum_{\hat{A}} \frac{N^2_j(\hat{A})}{\sigma^2_{ij}} \tag{18}$$

For this study, without explicit experimental covariance matrices, most of components of variance-covariance are considering null:

$$Cov(N_j(A); N_l(A')) = Var(N_j(A)).\delta_{AA'}.\delta_{jl} \quad \forall j, l \tag{19}$$

For n measurements of mass A, the mean normalized mass rate $\bar{R}(A)$ is equal to [18]:

$$\bar{R}(A) = W_1.R_1^i(A) + ... + W_n.R_n^i(A) = \left(\sum_{j=1}^{n} W_j.R_j^i(A) \right) \tag{20}$$

with:

$$W_{j=1,n} = \frac{C_{1j}^{-1} + ... + C_{nj}^{-1}}{C_{11}^{-1} + C_{12}^{-1} + ... + C_{1n}^{-1} + C_{21}^{-1} + ... + C_{2n}^{-1} + ... + C_{n1}^{-1} + ... + C_{nn}^{-1}}$$
$$= \left(\sum_{l}^{n} C_{lj}^{-1} \right) / \left(\sum_{l,j}^{n,n} C_{lj}^{-1} \right) \tag{21}$$

with $C_{lj} = Cov(R_l^i(A); R_j^i(A))$ for a same mass A and its variance is given by the following equation:

$$Var(\bar{R}(A)) = 1 / \left(\sum_{lj}^{n,n} C_{lj}^{-1} \right) \tag{22}$$

For n measurements of mass A and m measurements of mass A', the covariance of mean normalized mass rates is equal to:

$$Cov(\bar{R}(A); \bar{R}(A')) = \sum_{l,j=1}^{n,m} W_l \, C_{lj} \, W_j \tag{23}$$

with here $C_{lj} = Cov(R_l^i(A); R_j^i(A'))$ for two different masses A and A', thus $(n \times m)$ terms; and the weights:

$$W_{j=1,n} = \left(\sum_{l}^{n} C_{lj}^{-1} \right) / \left(\sum_{l,j}^{n,n} C_{lj}^{-1} \right) \quad and \quad W_{l=1,m} = \sum_{j}^{m} C_{lj}^{-1} / \sum_{l,j}^{m,m} C_{lj}^{-1} \tag{24}$$

with here $C_{lj} = Cov(R_l^i(A); R_j^i(A))$ for a same mass A with $(n \times n)$ covariance terms (or for same mass A' with $(m \times m)$ covariance terms)

References

[1] L. Fiorito et al. Annals of Nuclear Energy, **69**, (2014) 331-343.
[2] A. Chebboubi et al., EPJ Web of Conferences, **146**, (2017) 04021.
[3] K. H. Schmidt et al., Nucl. Data Sheets, **131**, (2016) 107-221.
[4] N. Terranova et al., Nucl. Data Sheets **95**, (2015) 225–230.
[5] D. Rochman et al., Annals of Nuclear Energy, **131**, (2016) 125-134.
[6] O. Leray et al.,EPJ Web of Conferences, **146**, (2017) 09023.
[7] EXFOR database, https://www-nds.iaea.org/exfor/exfor.htm
[8] W. J. Maeck et al., Allied Chem. Corp., Idaho Chem. Programs, **1142**, (1978) 09
[9] G. Diiorio and B. W. Wehring, Nuclear Instruments and Methods, **147**, (1977) 487 - 499.
[10] H. Thierens et al., Nuclear Instruments and Methods, **134**, (1976) 299 - 308.
[11] A. Bail, PhD Thesis, Université Sciences et Technologies - Bordeaux I, (2009).
[12] M.P. Tsoukatos et al., Dissertation Abstracts B (Sciences), volume 28, page 2759, 1968

[13] K.J. Rosman, Canadian Journal of Physics, volume 61, page 1490, 1983
[14] C.K. Mathiews, Physical Review, Part C, Nuclear Physics, volume 15, page 344, 1977/01
[15] B. Voirin, et al., EPJ N - Nuclear Sciences & Technologies, EDP Sciences, 2018, 4, pp.26
[16] S. Julien-Laferrière, et al., EPJ N - Nuclear Sciences & Technologies, EDP Sciences, 2018, 4, pp.25
[17] D.L. Smith, *Probability, statistics, and data uncertainties in nuclear science and technology, vol.4*, (OECD,Paris, 1991) 269p
[18] M. Schmelling, Phys. Scr. 51 676, 1995
[19] B. Diu, C. Guthmann, B. Roulet, D. Lederer, *Physique statistique* (Hermann, 1996) ISBN: 2-7056-6065-8
[20] The JEFF-3.1 nuclear data library, OECD, NEA, JEFF Report 21, ISBN 92-64-02314-3, (2006).
[21] M. B. Chadwick, et al., Nucl. Data Sheets **112** (2011) 2887.

The Los Alamos fission yield evaluation pipeline

M. R. Mumpower[1,*], *M.* Verriere[1], *A. E.* Lovell[1], *T.* Kawano[1], *I.* Stetcu[1], and *P.* Talou[2]

[1]Theoretical Division, Los Alamos National Laboratory, Los Alamos, NM, 87545, USA
[2]X-Computational Physics Division, Los Alamos National Laboratory, Los Alamos, NM, 87545, USA

Abstract. The nuclear data team at Los Alamos National Laboratory has undertaken a campaign to construct new fission yield evaluation. Significant advances have been made on a number of fronts. Nuclear potential energy surfaces can now be generated with the newly developed MicMac code based off the Finite-Range Liquid-Drop Model (FRLDM). This model can be incorporated into the Los Alamos de-excitation framework codes BeOH and CGMF to perform modeling of prompt, independent (IFY) and cumulative (CFY) fission yields that take into account prompt and beta-delayed neutrons and photons consistent with decay data. This is in stark contrast to what exists in evaluated nuclear data libraries today, where only a few incident energy points are used with limited physical insights and no consistency between IFY, CFY and decay data. We highlight the latest progress with application of neutron-induced fission of ^{235}U and ^{239}Pu.

1 Introduction

The understanding of fission, or splitting of a heavy nucleus into smaller fragments, is critical for applications of energy production, national security, and fundamental science [1, 2]. Los Alamos National Laboratory and partner institutions are currently undertaking a new Fission Product Yield (FPY) evaluation effort for use in such applications. The last major update to the US ENDF FPY evaluation came from the work of Chadwick *et al.* [3], nearly a decade ago. Since this time, remarkable advancement has been made in theoretical modeling of fission [4–7] and measurements campaigns have produced modern data over a range of excitation energies, see e.g. [8–10]. Our novel evaluation effort seeks to combine these recent advances, focusing on the description of FPY, to independent (IFY) and cumulative (CFY) fission yields that take into account prompt and delayed emission consistent with decay data [11, 12]. In what follows, we outline a typical workflow of the fission evaluation and report on the progress of theoretical modeling efforts.

2 The Fission Evaluation Pipeline

The current Los Alamos fission evaluation effort depends upon the construction of a state-of-the-art platform where many separate codes are combined together under one centralized Python3 framework, `NEXUS`. The purpose of the `NEXUS` framework is to handle the inputs and outputs between the various codes in discrete steps. This workflow allows for the parallel

*e-mail: mumpower@lanl.gov

development of models, codes and the inclusion of the latest data, providing incremental transactions between relevant steps. The output of a particular step may be tagged with relevant metadata, allowing for the ease of verification and validation. Modifications to data, inputs or outputs and the source code are preserved via `Git` version control for posterity and reproducibility. The branching and distributed nature of this version control system naturally mimics modern research environments where progress may be made in parallel on many different fronts and then combined at a later time into subsequent parts of an evaluation.

An example of a simple fission evaluation workflow is shown in Fig. 1. Primary fission fragment distributions may be fit to experimental mass yields with an appropriate fitting function. For major actinides, a superposition of five Gaussians has been used in past, and we find this to be suitable for the present work as well. Odd-even staggering in the charge yields can be included using the charge systematics Wahl [13]. In the case no experimental data are available, or when the predictive capabilities of theoretical models meet the required accuracy of evaluation, fragment yields may be generated using the Los Alamos `MicMac` code, see Sec. 3 for more details. Independent yields, which describe the fission products just after emission of prompt neutrons and γ-rays, may be constructed given the description of the fission fragment yield and information on the kinetic energy of the fragments. Monte Carlo based codes such as `CGMF` [14] and `FREYA` [15] provide access to individual fission events whereas deterministic methods have been employed in the `BeOH` code [16]. Deterministic methods are of particular importance to evaluation efforts as they probe both low and high statistical regions of the yield curve with equal computational cost. Cumulative yields, those that arise after all β-decays, and isomeric ratios can be tabulated by soliciting evaluated decay data. We also have another method at our disposal, the input of IFY into the `PRISM` reaction network code to obtain cumulative yields. We are working towards modifying `PRISM` to be used to calculate time-dependent applications such as reactor fuel burn up.

Parameter optimization for the various models and codes may happen intermittently over the course of the evaluation process. It is important to keep a detailed record of the particular inputs and outputs for each step in the evaluation. To achieve this in our workflow, we have implemented a special hashing algorithm which uses among other items, a unique universal identifier to track individual calculations. Once the fission yields have been thoroughly tested, official use evaluation files may be generated with the open source `DeCE` code.

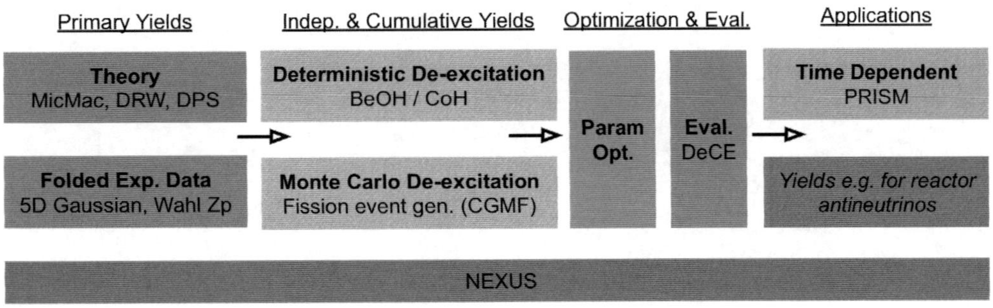

Figure 1. A schematic of some of the codes that comprise the Los Alamos fission evaluation pipeline. The workflow, indicated by arrows, is a combination of many distinct codes woven together by the Python3 framework, `NEXUS`. The creation of the evaluation file marks the beginning of public use for a variety of applications.

3 Fission Fragment Yields

Nuclear potential energy surfaces can now be generated with the newly developed `MicMac` code based off a Matched-Quadratic-Surace shape parameterization of the Finite-Range Liquid-Drop Model (FRLDM), see Dr. Verriere's first-author contribute to this conference proceedings. Several improvements have been made in comparison to older work [17], including an improved Strutinksy implementation and expanded axial harmonic oscillator basis that is extended from 12 to 20 shells. Further advances have been made in the self-consistent calculation of mass and charge yields. These quantities are directly calculated from the wave functions at the point of scission using a particle number projection technique [18]. Using this new method applied to fission fragments allows, for the first time, a prediction of odd-even staggering of charge yields that is seen in data and without invocation of systematics, for example, those of Ref. [13].

The calculation of the fission fragments themselves depends upon the pathways taken through the nuclear potential energy surface as discussed in Refs. [19, 20]. Figure shows the potential energy surfaces for (a) ^{236}U and (b) ^{240}Pu that are used in the calculation of ^{238}U(n,f) and ^{239}Pu(n,f). A large ridge (yellow) is seen in both nuclei for small to moderate mass asymmetries, leading to both of these systems fissioning asymmetrically. Using the theoretically calculated nuclear potential, the total kinetic energy of the fragments may also be produced when performing time-dependent calculations, such as with the Langevin approach [21]. Whether the desired accuracy can be achieved for evaluations with such an approach is a subject of current investigation.

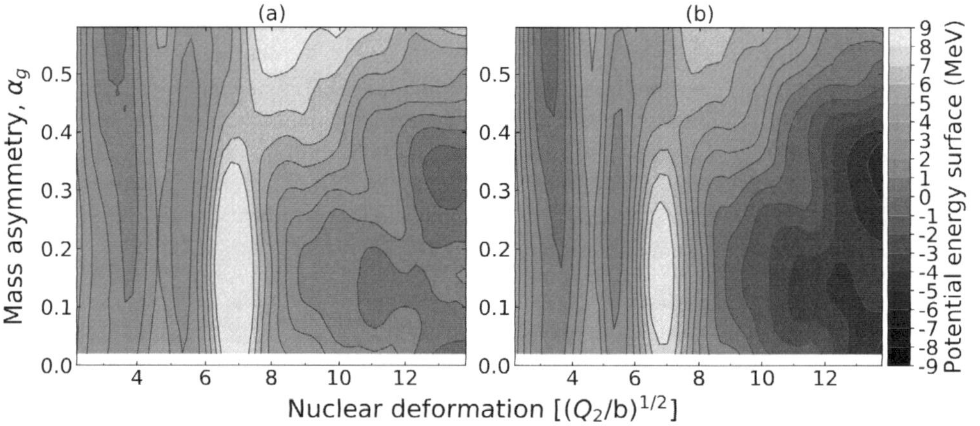

Figure 2. The calculated nuclear potential energy surfaces of (a) ^{236}U and (b) ^{240}Pu using the `MicMac` code recently developed at Los Alamos. The high-dimensional surfaces are projected down onto mass asymmetry and nuclear deformation. The yellow ridge seen at moderate elongations suggests both systems should fission asymmetrically.

4 Independent Yields

Due to the fast timescale of prompt particle emission, independent fission yields are the quantity often measured in experiments. This means that reconstruction of the primary fragments must ensue by invoking a model of the emission. The de-excitation of nascent fragments may be treated theoretically with statistical Hauser-Feshbach calculations [22]. Although there have been many studies performed using Monte Carlo methods, such as CGMF [4, 14, 23–25], the first attempt at this methodology using the deterministic code BeOH was recently implemented in Ref. [16] to study reactions with incident neutron energies below second-chance fission. The primary fission fragment yield and information on the total kinetic energy of the fragments are required for such calculations. The average excitation energy in each fragment may be approximated using anisothermal parameter that is defined as the ratio of the effective temperatures of the fragments. From this information, the statistical decay of the fragments may be followed, assuming compound nucleus formation, until all available excitation energy is exhausted.

The primary fragment and independent fission yields for ^{235}U(n,f) at thermal energies is shown in Fig. 3. The primary mass yield is fit to experimental data using the 5-Gaussian method and the Wahl charge systematics to produce the full yield, $Y(Z, A)$. Total kinetic energy as a function of mass number is also fit to experiment. With this method, the calculated independent fission yields reproduce the trends found in other evaluated libraries such as JENDL.

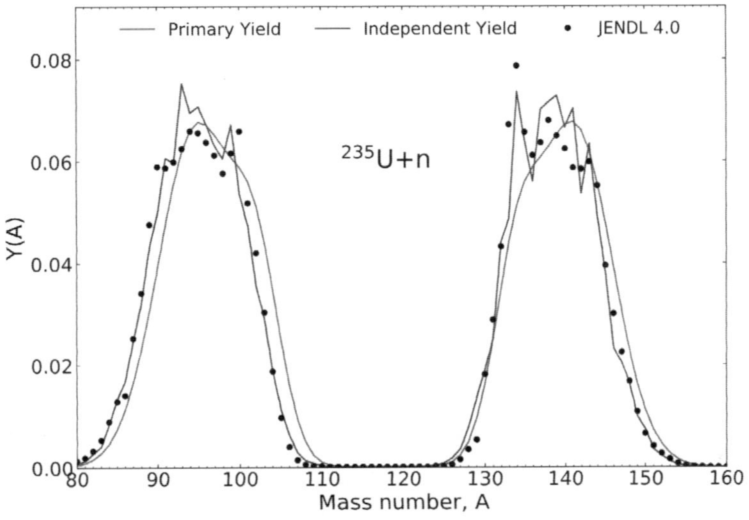

Figure 3. The calculated independent fission yield (blue) after statistical de-excitation of the nascent fragments (red) for the thermal neutron induced fission of ^{235}U. Calculations compared to JENDL-4 [26, 27].

This procedure also allows the outputting of the properties of emitted prompt neutrons and γ rays using well-tested statistical models [28] that are supplemented with evaluated and measured data when applicable [29]. Prompt neutron and γ-ray spectra associated with ^{235,238}U(n,f) and ^{239}Pu(n,f) are currently being studied. Good agreement with the experimental yields and isomeric ratios has been reported recently for the ^{235}U(n,f) reaction at thermal incident neutron energies [16]. An example of the calculated average number of prompt fission neutrons as a function of the mass of the fragment ca can be seen in Fig. 4 which qualitatively tracks measurements.

While CGMF already takes into account multi-chance fission for incident energies up to 20 MeV, work is in progress to extend this formalism beyond the second chance fission threshold using the deterministic method. Additional information is needed for this calculation, in particular the estimation of fission barriers [30]. This is because the transmission

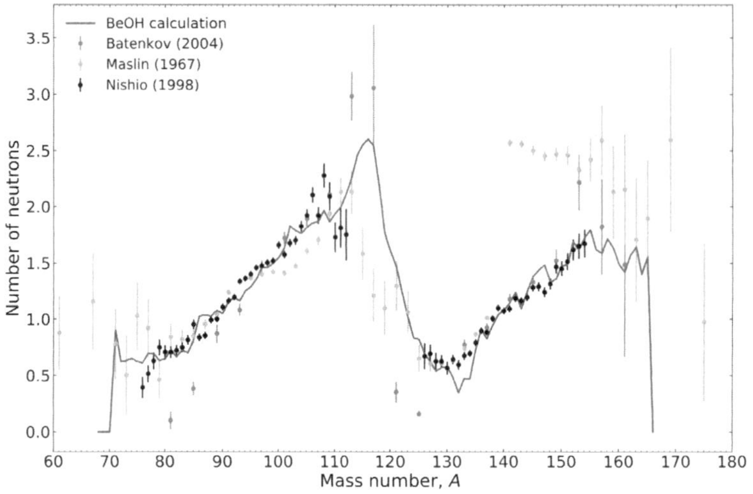

Figure 4. The calculated average number of prompt fission neutrons emitted (red) versus fragment mass number as compared to data (black and gray points) for the thermal neutron induced fission of ^{235}U. This calculation corresponds to the independent fission yield (blue line) in Fig. 3.

coefficient for the fission channel is calculated with penetration through a parabolic barrier as in Ref. [31]. The multi-nucleon transfer reaction data of [32] will be a critical test of such modeling capabilities.

5 Conclusions

Work on the Los Alamos fission evaluation pipeline has now begun in preparation for the DOE NA-22 program initiative to produce a new evaluation of fission product yields by 2023. The main goal of this work is to produce a unified framework, that will allow for the consistent evaluation of the fission fragment yields over a range of energies consistent with decay data. Several recent successes on the theoretical side have been reported in this proceedings. The development of the `MicMac` code for the creation of nuclear potential energy surfaces allows for the calculation of primary fission fragments in both charge and mass. The statistical de-excitation of these nascent fragments may be followed to produce both independent and cumulative yields. Active development on the modeling front is currently underway, with a focus on improving the energy dependence of fission yields.

6 Acknowledgements

We thank Mark Chadwick, Ramona Vogt, Jørgen Randrup and Peter Möller for valuable discussions. This work was supported by the US Department of Energy through the Los Alamos National Laboratory. Los Alamos National Laboratory is operated by Triad National Security, LLC, for the National Nuclear Security Administration of U.S. Department of Energy (Contract No. 89233218CNA000001).

References

[1] R. Capote, D.L. Smith, A. Trkov, EPJ Web of Conferences **8**, 04001 (2010)

[2] P. Talou, R. Vogt, J. Randrup, M.E. Rising, S.A. Pozzi, J. Verbeke, M.T. Andrews, S.D. Clarke, P. Jaffke, M. Jandel et al., The European Physical Journal A **54**, 9 (2018)

[3] M. Chadwick, T. Kawano, D. Barr, M.M. Innes, A. Kahler, T. Graves, H. Selby, C. Burns, W. Inkret, A. Keksis et al., Nuclear Data Sheets **111**, 2923 (2010), nuclear Reaction Data

[4] P. Talou, B. Becker, T. Kawano, M.B. Chadwick, Y. Danon, Phys. Rev. C **83**, 064612 (2011)

[5] R. Vogt, J. Randrup, Nuclear Data Sheets **118**, 220 (2014)

[6] T. Kawano, R. Capote, S. Hilaire, P. Chau Huu-Tai, Phys. Rev. C **94**, 014612 (2016)

[7] M.R. Mumpower, T. Kawano, P. Möller, Phys. Rev. C**94**, 064317 (2016), `1608.01956`

[8] L. Qi, M. Lebois, J.N. Wilson, A. Chatillon, S. Courtin, G. Fruet, G. Georgiev, D.G. Jenkins, B. Laurent, L. Le Meur et al., Phys. Rev. C **98**, 014612 (2018)

[9] N. Fotiades, P. Casoli, P. Jaffke, M. Devlin, R.O. Nelson, T. Granier, P. Talou, T. Ethvignot, Phys. Rev. C**99**, 024606 (2019)

[10] D. Ramos, M. Caamaño, A. Lemasson, M. Rejmund, L. Audouin, H. Álvarez-Pol, J.D. Frankland, B. Fernández-Domínguez, E. Galiana-Baldó, J. Piot et al., Phys. Rev. Lett. **123**, 092503 (2019)

[11] D. Brown, M. Chadwick, R. Capote, A. Kahler, A. Trkov, M. Herman, A. Sonzogni, Y. Danon, A. Carlson, M. Dunn et al., Nuclear Data Sheets **148**, 1 (2018), special Issue on Nuclear Reaction Data

[12] G. Audi, F.G. Kondev, M. Wang, W.J. Huang, S. Naimi, Chinese Physics C **41**, 030001 (2017)

[13] A.C. Wahl, Los Alamos Report (2002)

[14] B. Becker, P. Talou, T. Kawano, Y. Danon, I. Stetcu, Phys. Rev. C **87**, 014617 (2013)

[15] J. Verbeke, J. Randrup, R. Vogt, Computer Physics Communications **222**, 263 (2018)

[16] S. Okumura, T. Kawano, P. Jaffke, P. Talou, S. Chiba, Journal of Nuclear Science and Technology **55**, 1009 (2018), `https://doi.org/10.1080/00223131.2018.1467288`

[17] P. Möller, A.J. Sierk, T. Ichikawa, H. Sagawa, Atomic Data and Nuclear Data Tables **109**, 1 (2016), `1508.06294`

[18] M. Verriere, N. Schunck, T. Kawano, Phys. Rev. C**100**, 024612 (2019), `1811.05568`

[19] J. Randrup, P. Möller, Phys. Rev. Lett. **106**, 132503 (2011)

[20] J. Randrup, P. Möller, A.J. Sierk, Phys. Rev. C **84**, 034613 (2011)

[21] M.D. Usang, F.A. Ivanyuk, C. Ishizuka, S. Chiba, Phys. Rev. C**94**, 044602 (2016)

[22] W. Hauser, H. Feshbach, Phys. Rev. **87**, 366 (1952)

[23] I. Stetcu, P. Talou, T. Kawano, M. Jandel, Phys. Rev. C **90**, 024617 (2014)

[24] P. Jaffke, P. Möller, P. Talou, A.J. Sierk, Phys. Rev. C**97**, 034608 (2018), `1712.05511`

[25] A.E. Lovell, I. Stetcu, P. Talou, G. Rusev, M. Jandel, Phys. Rev. C **100**, 054610 (2019)

[26] K. Shibata, O. Iwamoto, T. Nakagawa, N. Iwamoto, A. Ichihara, S. Kunieda, S. Chiba, K. Furutaka, N. Otuka, T. Ohsawa et al., Journal of Nuclear Science and Technology **48**, 1 (2011), `https://doi.org/10.1080/18811248.2011.9711675`

[27] K. Shibata, O. Iwamoto, T. Nakagawa, N. Iwamoto, A. Ichihara, S. Kunieda, S. Chiba, J. Katakura, N. Otuka, Journal of Korean Physical Society **59**, 1046 (2011)

[28] T. Kawano, S. Chiba, H. Koura, Journal of Nuclear Science and Technology **43**, 1 (2006)

[29] M.R. Mumpower, T. Kawano, J.L. Ullmann, M. Krtička, T.M. Sprouse, Phys. Rev. C**96**, 024612 (2017), `1706.07504`

[30] P. Möller, A.J. Sierk, T. Ichikawa, A. Iwamoto, M. Mumpower, Phys. Rev. C**91**, 024310 (2015)

[31] D.L. Hill, J.A. Wheeler, Physical Review **89**, 1102 (1953)

[32] K. Hirose, K. Nishio, S. Tanaka, R. Léguillon, H. Makii, I. Nishinaka, R. Orlandi, K. Tsukada, J. Smallcombe, M.J. Vermeulen et al., Phys. Rev. Lett. **119**, 222501 (2017)

Quantified uncertainties in fission yields from machine learning

A.E. Lovell[1,*], *A.T.* Mohan[1], *P.* Talou[1], and *M.* Chertkov[1,2]

[1]Los Alamos National Laboratory, Los Alamos, New Mexico, 87545, USA
[2]University of Arizona, Tucson, Arizona, 85721, USA

Abstract. As machine learning methods gain traction in the nuclear physics community, especially those methods that aim to propagate uncertainties to unmeasured quantities, it is important to understand how the uncertainty in the training data - coming either from theory or experiment - propagates to the uncertainty in the predicted values. Gaussian Processes and Bayesian Neural Networks are being more and more widely used, in particular to extrapolate beyond measured data. However, studies are typically not performed on the impact of the experimental errors on these extrapolated values. In this work, we focus on understanding how uncertainties propagate from input to prediction when using machine learning methods. We use a Mixture Density Network (MDN) to incorporate experimental error into the training of the network and construct uncertainties for the associated predicted quantities. Systematically, we study the effect of the size of the experimental error, both on the reproduced training data and extrapolated predictions for fission yields of actinides.

1 Introduction

Post-scission fission fragment mass yields are a necessary input for a variety of applications, including energy and stockpile stewardship, fission decay codes such as `CGMF` [1, 2], `FREYA` [3, 4], and `BeoH` [5], and the synthesis of nuclei in astrophysical environments, e.g., [6]. Even for applications that only require fission yields for major actinides, experimental data are sorely lacking for yields as a function of mass, charge, and fragment kinetic energy for all reactions and incident energies. The challenges are compounded when inputs are required for astrophysical calculations, where fission yields across the nuclear chart are required, in particular out toward the neutron dripline where measurements are currently impossible. Here, theoretical predictions or systematics derived from data are needed to make these calculations possible.

There has been much progress recently in calculating mass yields from both microscopic [7, 8] and microscopic-macroscopic models [9, 10]. Although these methods are very promising, these calculations are not yet at the precision required for many applications. As a complement, systematics have been developed by fitting to experimental data [11–13]. In addition, machine learning techniques, mainly the Bayesian Neural Network, have recently begun to be used to make predictions for mass yields [14]. Here, we continue along those

*e-mail: lovell@lanl.gov

lines, proposing to use a rather novel machine learning algorithm, the Mixture Density Network [15], to fill in the sparse set of fission yield data with quantified uncertainties.

The outline of this work is as follows: in Section 2, we briefly discuss the Mixture Density Network and the numerical details that are needed for the algorithm, results are presented and discussed in Section 3, and we conclude in Section 4.

2 Mixture Density Networks

To fill in the sparse landscape of nuclear data, we propose using a rather novel machine learning technique, the Mixture Density Network (MDN) [15]. While standard feed-forward Neural Networks (NN) aim to map input, x, directly to output, y, through a non-linear function, $y = f(x)$, the MDN describes the output as a mixture of Gaussian functions,

$$f(x) = \sum_{i=1}^{N} = \alpha_i(x)\mathcal{N}(\mu_i(x), \sigma_i(x)). \tag{1}$$

The neural network is then used to determine the weights, means, and standard deviations of each Gaussian mixture, α_i, μ_i, and σ_i. In contrast to standard deterministic NNs where predictions typically consist of a mean value and possibly a standard deviation (leaving it to the user to assume a shape for the posterior distribution), the MDN predicts the full posterior distribution for each output. Through the construction of the full posterior, we are able to fully quantify the uncertainty on each predicted value. A mean and standard deviation can be calculated, but we still have access to the exact shape of the distribution and any correlations between the predicted outputs.

2.1 Numerics for the MDN

For our convergence studies, we use a single Gaussian mixture. To determine the mean, width, and standard deviation of this Gaussian, the NN has four layers with ten nodes per layer. The training set consists of 3000 $[A, Y(A)]$ pairs where $Y(A)$ is sampled from a Gaussian distribution centered on a fit to experimental data with a predefined standard deviation. (The convergence of the number of points in the training set is shown in Sec. 3.1.) For the testing set, we randomly sample 10000 mass values in the range $[A_c/2 - 50, A_c/2 + 50]$, where A_c is the mass of the compound nucleus. Algorithms, training sets, or testing sets that deviate from these specifications are detailed in the text.

3 Results

In this proceedings, we discuss briefly the convergence studies performed for the MDN, focusing on how the uncertainties propagate from the training set to the MDN prediction. We then show results of training on a set constructed from experimental data, and finally discuss the potential of the MDN for interpolating between training sets.

3.1 Convergence studies

First, we test the convergence of the number of points needed in the training set using mass yields for ^{252}Cf spontaneous fission. To do this, for each mass in the range $[A_c/2 - 50, A_c/2 + 50]$, we randomly sample the mean of $Y(A)$ from a Gaussian with a standard deviation of 20% of the mean. While a 20% error is much larger than what is typically measured for the

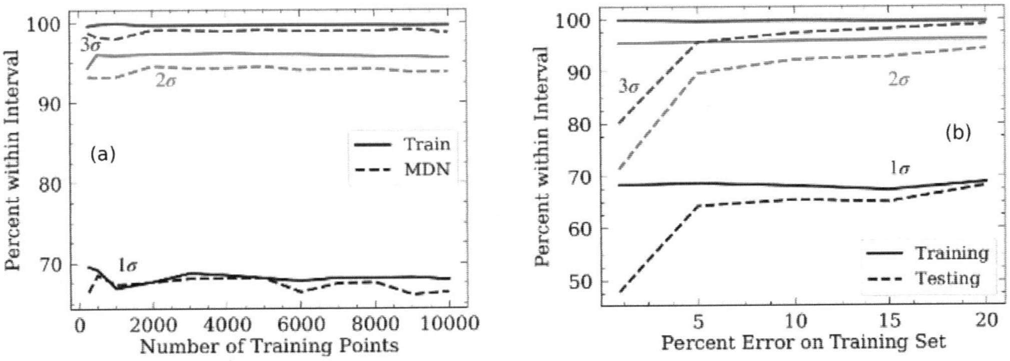

Figure 1. Percentage of the training set (solid lines) and MDN validation (dashed lines) that fall within 1σ (black), 2σ (red), and 3σ (blue) of the mean value for (a) the number of samples in the training set and (b) percentage error on mass yields used to construct the training set of 3000 samples.

mass yields, if the convergence holds for this large uncertainty, then the training set should be more than large enough for smaller values of the uncertainty as well.

For each convergence study, we compare the training set and the MDN prediction by showing the percentage of the training samples and MDN pulls that fall within 1, 2, and 3 standard deviations of the known mean value (e.g. the value sampled to construct the training set). If the posterior distribution of the MDN is the same as the distribution of the training set, ~68% of samples should fall within 1σ, 95% within 2σ, and 99% within 3σ. This should also be reflected in the distribution of the training set, as long as there is an adequate number of samples in the training set.

In Fig. 1(a), we show the convergence of the percentage of training samples and MDN predictions that fall within 1, 2, and 3 σ as a function of the number of samples in the training set. We see that even for the training set, below about 1000 samples, the statistics are not enough to consistently construct these intervals. For the MDN validation, around 3000 samples, the predictions have mostly converged, although we note that the 1σ intervals are less stable (flat) than the other two. In each case, we see that a slightly lower percentage of MDN predicted values of $Y(A)$ fall within each standard deviation than are within the training set, meaning that the MDN somewhat increases the uncertainty during the propagation.

In Fig. 1(b), we show the percentage within each standard deviation as a function of the error that was assumed on the training set. Immediately, we notice that the uncertainties are propagated poorly when a 1% error on the training set is taken, as uncertainties here are much larger than the errors included on the training set. It is also worth noting that, in this case, we needed four Gaussian mixtures for the neural network to converge. Our recent studies show that this is due to orders of magnitude spanned by the mass yields, and when the tails of the distribution are removed, only one Gaussian is needed in the mixture. For the larger values of the uncertainty on the training set (5%, 10%, 15%, and 20%), the MDN still produces uncertainties that are slightly larger than what was put on the training set but not as significantly as with the 1% errors.

3.2 Reproducing experimental data

We next show how the MDN can constrain several data sets simultaneously, using experimentally measured mass yields for ^{252}Cf(sf) [16–19]. To construct the training set, we draw

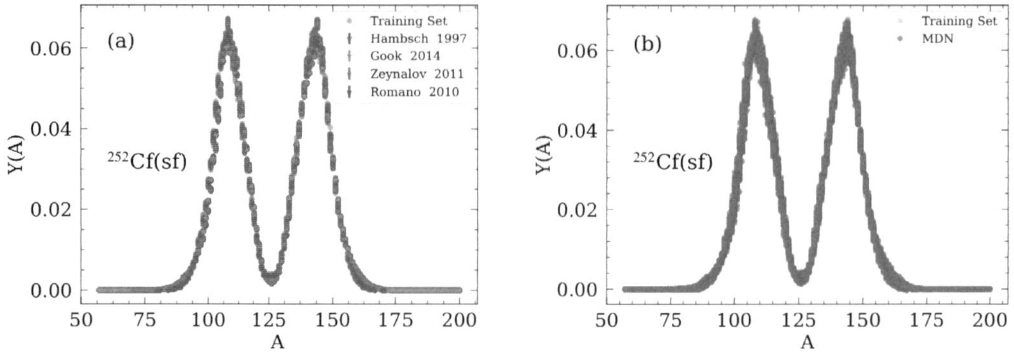

Figure 2. (a) Comparison between experimental mass yields for ^{252}Cf(sf) from [16–19] (colored stars) and the constructed training set for the MDN (purple circles). (b) MDN prediction (red) underlaid behind the training set (purple).

100 samples for each mass value from a Gaussian centered at the experimentally measured yield with a standard deviation equal to the reported uncertainty. Each mass is assumed to be independent from the others within the same experiment, and we assume zero correlations between the four experiments. This assumption is a poor one: we know that there are correlations within and between experiments. Studies to include these correlations are underway.

In Fig. 2(a), we show the training set sampled from the experimental mass yield data compared to the experimental data. Although ^{252}Cf(sf) mass yields have been measured very precisely over the last several decades, some small discrepancies between the four data sets can still be seen, in particular at the peaks of the distribution and in the symmetric region. This gives us an opportunity to investigate how the MDN handles overlapping but not identical data sets - without having to struggle with completely discrepant data.

Here, we again use a four-layer neural network with ten nodes per layer. In addition, we choose a mixture of four Gaussians to allow for the possibility that each data set will need to be described by a separate Gaussian. In panel (b) of Fig. 2, we then shown the resulting MDN prediction in red compared to the training set in purple. The MDN is qualitatively able to reproduce the training set, including the spread of the data (for instance around $A \approx 160$, the peaks of the distribution, and the symmetric mass region).

We can construct the mean value and standard deviation of $Y(A)$ from the MDN prediction. The mean values fall within the spread of the experimental data, and although the standard deviation is larger than the experimental errors, preliminary studies show that the spread can be reduced by introducing physical constraints into the training set (e.g. enforcing the symmetry between the light and heavy peaks). Additionally, a study of the posterior distribution of the MDN predictions shows that the MDN is not simply reproducing the distribution of the training set but gives some insight into the credibility of each data set.

To give more detail about the distribution of the MDN, we can also look at the mean yield values for each mass and the spread of the MDN prediction. These fall within the spread of the data, and although the resulting standard deviation of the MDN is larger than the experimental errors on the data, it reflects the spread in the data. In addition, looking at the distribution of the MDN predictions for specific values of the mass, we could see that the MDN doesn't just average over the four data sets, but it gives a distribution based on the strength of each set.

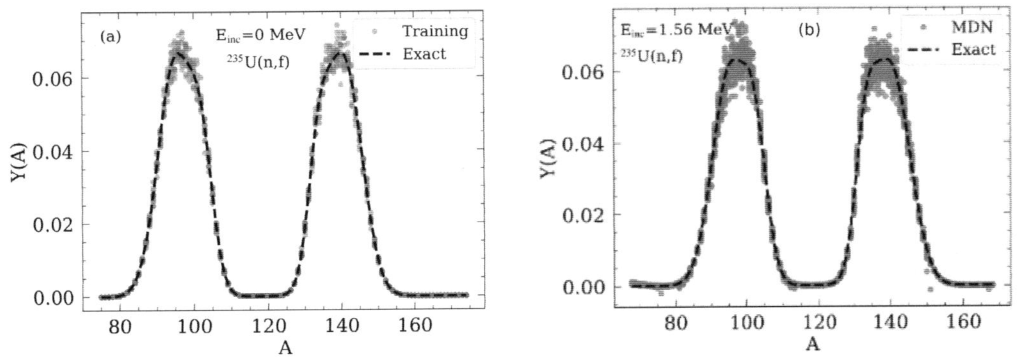

Figure 3. Mass yields for ^{235}U(n,f). (a) Training set at $E_{\rm inc}$ = 0 MeV (thermal) compared to the exact parameterization from CGMF and (b) MDN prediction for $E_{\rm inc}$ = 1.56 MeV compared to the exact CGMF parameterization.

3.3 Interpolating the energy dependence

While being able to reproduce the training set is a good test of the MDN, ultimately to be useful, we need to be able to interpolate between the training sets. To do this, we use mass yields for ^{235}U(n,f) as a function of incident neutron energy. We train on whole number energies from 0 (to denote thermal) to 10 MeV, and interpolate between energies on this grid. Here, we assume a 5% uncertainty for each mass yield and take the mean values to be the energy-dependent fit to experimental data. By default, this parameterization does not take into account multi-chance fission.

In Fig. 3(a), we show one example of the training set compared to the exact parameterization from CGMF for incident thermal neutrons. Instead of directly mapping each mass to a yield, the MDN maps pairs of inputs (mass and incident neutron energy) to a single output (mass yield). To test the ability of the MDN to interpolate between the energies in the training set, we randomly sample real numbers in the energy range of 0 to 10 MeV. In Fig. 3(b), we show one example from the testing set at an incident neutron energy of 1.56 MeV. The MDN prediction is again compared to the exact energy-dependent parameterization, in black. Although this energy is nearly exactly in the middle of two of the training points (1 MeV and 2 MeV), the MDN prediction is still centered around the exact parameterization. The spread in the MDN prediction is about twice as large as the 5% uncertainty included on the training set; preliminary studies indicate that the large uncertainties are driven by the tails of the distributions.

4 Conclusion

In summary, we are using a novel machine learning technique, the Mixture Density Network, to reproduce fission fragment mass yields with well-quantified uncertainties. We test the convergence of the MDN as a function of the number of data points needed in the training set and confirm that the MDN can reproduce mass yields for the spontaneous fission of ^{252}Cf with a range of uncertainties from 1% to 20%. We also use the MDN to simultaneously train on four different sets of experimental data for the mass yields of ^{252}Cf(sf) to provide a mean and posterior distribution for each mass value. In addition, we show how then MDN can be used to interpolate between training sets using energy-dependent yields for ^{235}U(n,f). In each case, the MDN slightly increases the uncertainty when propagating errors from training set to prediction, performing worse for smaller errors.

Many of these features of the MDN are still under investigation, including why the uncertainties are overpredicted for small errors on the training set and why more Gaussian mixtures are needed for the stability of the network when small ($\sim$1%) errors are included in the training set. Preliminary studies suggest that this is due to the several orders of magnitude covered by the fission yields in the tails of the distributions; if this is the case, this challenge can be mitigated by removing this part of the distribution during training or by using a logarithmic scaling. After seeing the ability of the MDN to interpolate between training sets, we can next investigate the power of the MDN to extrapolate beyond the given training set. Once the numerics are stable and we have a clear understanding of how well the MDN can extrapolate, this method will be extended to make predictions across a region of the nuclear chart. Efforts along all of these paths are currently underway.

This work was performed under the auspice of the U.S. Department of Energy by Los Alamos National Laboratory under Contract 89233218CNA000001 and was supported by the Office of Defense Nuclear Nonproliferation Research & Development (DNN R&D), National Nuclear Security Administration, U.S. Department of Energy. We gratefully acknowledge the support of the U.S. Department of Energy through the LANL/LDRD Program and the Center for Non Linear Studies.

References

[1] P. Talou, T. Kawano, I. Stetcu, P. Jaffke, M.E. Rising, A.E. Lovell, Comp. Phys. Comm. (in progress)
[2] B. Becker, P. Talou, T. Kawano, Y. Danon, I. Stetcu, Phys. Rev. C **87**, 014617 (2013)
[3] J. Verbeke, J. Randrup, R. Vogt, Computer Physics Communications **191**, 178 (2015)
[4] J. Verbeke, J. Randrup, R. Vogt, Computer Physics Communications **222**, 263 (2018)
[5] S. Okumura, T. Kawano, P. Talou, P. Jaffke, S. Chiba, J. Nucl. Sci. Tech. **55**, 1009 (2018)
[6] N. Vassh, R. Vogt, R. Surman, J. Randrup, T.M. Sprouse, M.R. Mumpower, P. Jaffke, D. Shaw, E.M. Holmbeck, Y. Zhu et al., Journal of Physics G: Nuclear and Particle Physics **46**, 065202 (2019)
[7] N. Schunck, D. Duke, H. Carr, A. Knoll, Phys. Rev. C **90**, 054305 (2014)
[8] D. Regnier, N. Dubray, N. Schunck, M. Verrière, Phys. Rev. C **93**, 054611 (2016)
[9] A.J. Sierk, Phys. Rev. C **96**, 034603 (2017)
[10] P. Möller, T. Ichikawa, Eur. Phys. J. A **51**, 173 (2015)
[11] A.C. Wahl, Tech. rep., Los Alamos Nat. Lab. LA-13928 (2002)
[12] U. Brosa, S. Grossmann, A. Müller, Physics Reports **197**, 167 (1990)
[13] K.H. Schmidt, B. Jurado, C. Amouroux, Tech. Rep. NEA/DB/DOC(2014)1, Nuclear Energy Agency (2014)
[14] Z.A. Wang, J. Pei, Y. Liu, Y. Qiang, Phys. Rev. Lett. **123**, 122501 (2019)
[15] C.M. Bishop, Tech. rep., Aston University, Department of Computer Science and Applied Mathematics (1994)
[16] A. Göök, F.J. Hambsch, M. Vidali, Phys. Rev. C **90**, 064611 (2014)
[17] C. Budtz-Jørgensen, H.H. Knitter, Nuclear Physics A **490**, 307 (1988)
[18] C. Romano, Y. Danon, R. Block, J. Thompson, E. Blain, E. Bond, Phys. Rev. C **81**, 014607 (2010)
[19] S. Zeynalov, F.J. Hambsch, S. Obertstedt, J. Korean Phy. Soc. **59**, 1369 (2011)

Toward next JENDL Fission Yield Data and Decay Data

Futoshi Minato[1,*]

[1]Nuclear Data Center, Japan Atomic Energy Agency, Tokai, Ibaraki 319-1195, Japan

Abstract. JENDL group has developed evaluated fission yield and decay data. They are now planned to be upgraded in order to embody new experimental data and theoretical developments obtained after the last release. In this paper, we briefly describe a plan and present activity toward next JENDL fission yield and decay data.

1 Introduction

Japanese Evaluated Nuclear Data Library (JENDL) keeps various kinds of evaluated nuclear data and contributes developments of nuclear physics and its applications. In the special purpose files, JENDL FP ("F"ission "P"roduct) Fission Yield Data file 2011 (JENDL/FPY-2011) and JENDL FP Decay Data file 2011 (JENDL/FPD-2011) were released in 2011 [1]. They were consistently evaluated to reproduce decay heat generated from FP after thermal, fast, and 14 MeV neutron induced fission for uranium, plutonium and thorium, using up-to-date knowledge of nuclear physics at that time. Later on, in order to complement other nuclei than FP, JENDL Decay Data File 2015 (JENDL/DDF-2015) was released in 2015 [2]. The number of nuclei compiled in JENDL/FPD-2011 is 1,284, while JENDL/DDF-2015 grows into 3,287.

After the release of those evaluated nuclear data files, some issues were found in the JENDL/FPY-2011. JENDL/FPY-2011 was originally produced on the basis of England and Rider evaluation [3]. However, it was found that fission yield of ^{86}Ge in thermal neutron fission of ^{235}U was incorrectly cited from an original experimental data and an extraordinary large yield has been assigned for ^{86}Ge in England and Rider evaluation [3]. Although this problem had been already pointed out in 2002 by Oyamatsu [4], no special treatment was done when JENDL/FPY-2011 was made. This problem as well as other extraordinary large fission yields of thermal neutron induced fission of ^{235}U was corrected by using a theoretical model in the revised version of JENDL/FPY-2011 released in 2016 [5]. Simultaneously, we realize an importance of re-evaluation of fission yield and decay data because a long time has passed since England and Rider evaluation [3] was published.

There still exist issues that are not solved in the present evaluated fission yield and decay data. The first is about the summation calculation of delayed neutron yields using JENDL/FPY-2011 and JENDL/FPD-2011 (or JENDL/DDF-2015). While we paid a special attention to the summation calculations of decay heat when JENDL/FPY-2011 and JENDL/FPD-2011 were evaluated, we did not put effort in those of delayed neutron yields. As a result that we ignored delayed neutron yields in the evaluation, large deviations were

*e-mail: minato.futoshi@jaea.go.jp

left between the summation calculation by JENDL/FPY-2011 and JENDL/FPD-2011 and experimental data (this will be shown in sec. 2). Due to the correction of ^{86}Ge in the revision of JENDL/FPY-2011 [5], the summation calculation of the delayed neutron yields for ^{235}U was somewhat improved. However, there still remain gaps between the summation calculation and experimental data not only for ^{235}U but also other nuclei. In addition, as a specific problem in the JENDL decay data, delayed neutron spectrum is not compiled in JENDL/FPD-2011 and JENDL/DDF-2015, although it is included in ENDF/VIII.0 decay data. It is discussed that delayed neutron spectra can be used for nuclear non-destructive analysis [6]. This lack reduces potential possibilities of applications using JENDL decay data.

Besides delayed neutron yields and decay heats, a lot of observables are resulted from nuclear fission and subsequent decay of FP, for example, prompt fission γ emission, delayed γ, β, and neutrino emissions. Those observables are radioactive until all fission fragment turns into stable nuclei. It is widely discussed in recent years that those observables can be applied to non-destructive analysis, monitoring nuclear reactor, etc. [7–11]. The past JENDL fission yield and decay data, however, have not paid a particular attention to those observables as well as delayed neutron yields. Next JENDL evaluated fission yield and decay data should consider those observables and be evaluated consistently and simultaneously in order to provide a more reliable evaluated data.

Since the release of JENDL/FPD-2011 and JENDL/DDF-2011, new experimental data have been obtained. In particular, half-lives and delayed neutron branching ratios have been actively measured in the last decade due to developments of radio isotope beam facilities in the world. Moreover, the total absorption gamma-ray spectroscopy (TAGS) data has been also newly measured since 2011. The TAGS data has played an important role to improve the summation calculation of decay heats when JENDL/FPY-2011 and JENDL/FPD-2011 were evaluated. We should consider those new experimental data in next JENDL evaluated fission yield and decay data.

Even though there are a plenty experimental data available, all of information on decay data is not necessarily known. To produce a completely arranged decay data file, theoretical models are used to compensate the lacks of experimental data. Gross theory and FRDM+QRPA models have been applied to JENDL and ENDF decay data, respectively. For next JENDL decay data file, we plan to make use of a microscopic model of in addition to the Gross theory used for JENDL/FPD-2011.

In this paper, we summarize a plan of next JENDL evaluated fission yield and decay data. In sec. 2.1, we describe new points to be included in the next JENDL decay data. In sec. 2.3, errors found in JENDL/FPD-2011 and JENDL/DDF-2015, which are to be corrected in next JENDL decay data are listed In sec. 2.2, two microscopic theoretical models that we are developing so far are explained and a recent activity using them is given. In sec. 3, a plan of new JENDL fission product yield data is briefly described. In sec. 4, we conclude our remarks.

2 Improvements planned in next JENDL decay data

In this section, we describe a main feature of next JENDL decay data. In sec. 2.1, new points to be included in next JENDL decay data are briefly described. In sec. 2.2, we explain microscopic theoretical models that are planned to be used for next JENDL decay data. In sec. 2.3, errors found in JENDL/FPD-2011 and JENDL/DDF-2015 are listed. Those will be corrected in the next JENDL decay data.

2.1 New points from the previous data

We list the new points to be included in the next JENDL decay data in the following:

1. Numerical values in the decay data will be updated by the latest version of ENSDF and experimental data including newly measured TAGS data.
2. Spectra of β-ray, delayed neutron, and other light particles will be included.
3. New theoretical calculations will be applied for short-life nuclei where no experimental data exist.
4. The decay data will be benchmarked by aggregated data such as decay heat, delayed neutron yield, prompt fission gamma spectra, etc.
5. Light exotic nuclei such as ^{4,5,6}H, which have been omitted in JENDL/DDF-2015, will be compiled.

As mentioned in sec.1, delayed neutron yields were not used for the benchmark test of JENDL/FPY-2011 and JENDL/FPD-2011, although a special attention was put in decay heats. Figure 1 and 2 show the delayed neutron activities as a function of time for thermal and fast neutron induced fission, respectively. The summation calculation by using JENDL/FPY-2011 and JENDL/FPD-2011 is compared with Keepin's six group evaluation [12]. As seen in the figure, the summation calculations of JENDL/FPY-2011 and JENDL/FPD-2011 systematically overestimate the delayed neutron activities of Keepin's evaluations both for thermal and fast neutron induced fission in a wide range of time. We plan to improve the shortcoming of the summation calculation of delayed neutron yields as well as other observables in the next JENDL evaluated fission yield and decay data.

2.2 Theoretical Model Applications

As mentioned in sec. 1, theoretical models are applied in evaluating decay data where no experimental data are available. We have been developing several theoretical models to apply them to nuclear decay data. Gross theory [13–15] is one of our powerful approaches. This model gives a reasonable agreement with experimental data in half-lives, delayed neutron branching ratios, and average β and γ emission energies in a wide range of nuclei.

Another promising model is a quasiparticle random-phase-approximation (QRPA) on the basis of non-relativistic Skyrme energy density functional (Skyrme EDF) [16] and covariant density functional [17]. We have calculated half-lives and delayed neutron branching ratios of FPs using QRPA+Skyrme EDF [16]. We developed the model considering a finite-range proton-neutron isospin $T = 0$ force. The delayed neutron branching ratio and its spectra are calculated with Hauser-Feshbach statistical model. We also incorporate the statistical model to QRPA based on the covariant density functional [17] to estimate delayed neutron branching ratio in a more realistic way. At present, we are making decay data table including half-lives, delayed neutron and fission branching ratios, average β and γ emission energies, emitted light particle $(e^{\pm}, \bar{\nu}_e, n, \alpha)$ spectra, using QRPA approaches described above. The result will be submitted elsewhere.

To select the best model, results of theoretical approaches have to be compared with available experimental data. Half-lives are often used as the benchmark of theoretical models. However, we plan to take a more careful test for theoretical models by using experimental data of delayed neutron branching ratios, and its activities, emitted particle energies, and etc. We have studied the summation calculation of decay heat using JENDL/FPY-2011 and decay

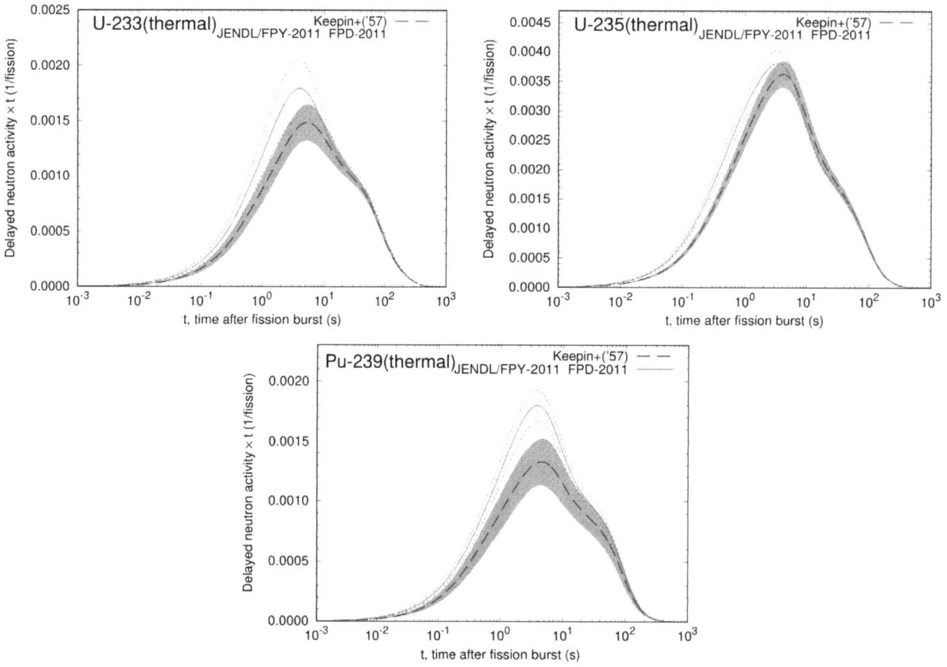

Figure 1. Delayed neutron activities for thermal neutron induced fission for $^{233,235}U$ and ^{239}Pu multiplied by time (t). In the panels, the results of Keepin's six group evaluation [12] (the dashed line) and the summation calculation by using JENDL/FPY-2011 and JENDL/FPD-2011 (the solid line) are shown. Errors of Keepin's six group are shown by the shaded area, while the uncertainties of the summation calculation are shown with the dotted lines.

data provided purely by the above theoretical models. We found that those tests are also useful to see the best model for decay data. The result of benchmark of theoretical models will be also submitted elsewhere.

2.3 Missing and mistaken data found in JENDL/FPD-2011

We recognized several errors in JENDL/FPD-2011 and JENDL/DDF-2015, which are listed in the following. Those will be corrected in the next JENDL decay data. The corrected version of JENDL/FPD-2011 and JENDL/DDF-2015 are also planned to be distributed.

1. γ-ray spectra of ^{241}Am are wrong in JENDL/DDF-2015.
2. An isomeric state of ^{235}U is populated by α decay of ^{239}Pu. However, the isomeric state of ^{235}U is missing in JENDL/DDF-2015.
3. Double β^--decay (RTYP=1.1) is assigned for neutron deficient nucleus Kr-78. The correct decay type is double electron capture (RTYP=2.2).
4. Two ground states are given for ^{32}Na, ^{93}Ru, ^{101}Pd, ^{131}Pr, ^{137}Pm, and ^{183}Ir. One of them must be an isomeric state.
5. FPs compiled in JENDL/FPY-2011 should be in JENDL/FPD-2015 as well. However, it is found that isomeric states of ^{84m}As, ^{109m}Ru, and ^{143m}Xe are not in JENDL/FPD-2011.

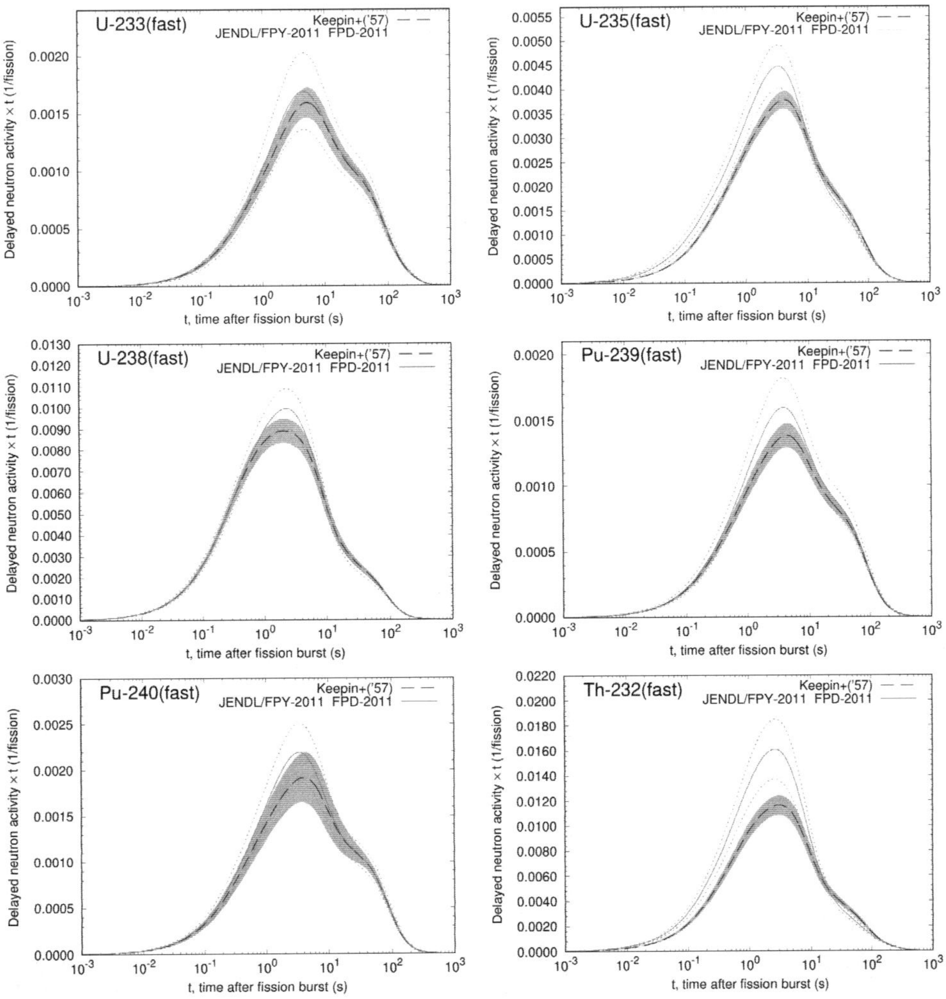

Figure 2. Same as Fig. 1, but for delayed neutron activities for fast neutron induced fission for $^{233,235,238}U$, $^{239,240}Pu$, and ^{232}Th.

6. Decay data of FPs in JENDL/FPD-2011 and JENDL/DDF-2015 should be identical. However, it is found that ^{82}Se, $^{83,85}Sr$, ^{96}Zr, ^{100}Mo, ^{111m}Pd, ^{116}Cd, $^{128,130}Te$, $^{134,136}Xe$, $^{136,138,142}Ce$, ^{150}Nd, and ^{151}Eu are not identical.

7. ^{109}Cd decays to an isomeric state of ^{109}Ag. However, it decays to the ground state in JENDL/FPD-2011 and JENDL/DDF-2015.

3 Next JENDL Fission Yield

We briefly describe an activity for new fission product yield data in this section. It is planned to adopt the evaluated fission yields produced by Tokyo institute of Technology group [18] as JENDL-5 fission yield data file. The evaluated data have several developments as compared to JENDL/FPY-2011. For example, it considers up-to-date insights on nuclear physics and is benchmarked by product inventory (PI) data provided from various nuclear reactors. In particular, it includes covariance data, thus it enables us to estimate a more reliable uncertainties when summation calculations are performed.

4 summary

We described a plan and recent activity for next JENDL evaluated fission yield and decay data. We are currently working to distribute them within a few years. Other than described in this paper, any request to our next evaluated data is welcomed.

References

[1] J. Katakura, *JENDL FP Decay Data File 2011 and Fission Yields Data File 2011*, JAEA-Data/Code 2011-025 (2011)

[2] J. Katakura, F. Minato, *JENDL Decay Data File 2015*, JAEA-Data/Code 2015-030 (2016)

[3] T.R. England, B.F. Rider, *Evaluation and Compilation of Fission Product Yields*, LA-UR-94-3106, ENDF-349 (1994)

[4] K. Oyamatsu, Journal of Nuclear Science and Technology **39**, 1109 (2002)

[5] J. Katakura, F. Minato, K. Ohgama, EPJ Web of Conferences **111**, 08004 (2016)

[6] T. Akyurek, S. Usman, Progress in Nuclear Energy **85**, 525 (2015)

[7] L.E. Smith, E.K. Mace, A.C. Misner, M.W. Shaver, IEEE Transactions on Nuclear Science **57**, 2247 (2010)

[8] P.H. Liem, S. Amini, A.G. Hutagaol, T.M. Sembiring, Annals of Nuclear Energy **56**, 57 (2013)

[9] M.V. Mora, A.G. Padilla, J.L.C. Palomino, L.A.A. Terremoto, Progress in Nuclear Energy **53**, 344 (2011)

[10] H. Furuta, Y. Fukuda, T. Hara, T. Haruna, N. Ishihara, M. Ishitsuka, C. Ito, M. Katsumata, T. Kawasaki, T. Konno et al., Nuclear Instruments and Methods in Physics Research Section A: Accelerators, Spectrometers, Detectors and Associated Equipment **662**, 90 (2012)

[11] F. Suekane, Nuclear Physics B - Proceedings Supplements **235-236**, 33 (2013), the XXV International Conference on Neutrino Physics and Astrophysics

[12] G.R. Keepin, T.F. Wimett, R.K. Zeigler, Phys. Rev. **107**, 1044 (1957)

[13] M. de Saint Simon, O. Sorlin, eds., *Exotic nuclei and atomic masses. Proceedings, International Conference, ENAM'95, Arles, France, June 19-23* (1995)

[14] T. Yoshida, T. Tachibana, J. Nucl. Sci. Technol. **37**, 491 (2000)

[15] T. Tachibana, RIKEN Review, Focused on Models and Theories of the Nuclear Mass **26**, 109 (2000)

[16] F. Minato, EPJ Web of Conferences **122**, 10001 (2016)

[17] T. Marketin, L. Huther, G. Martinez-Pinedo, Phys. Rev. C **93** (2016)

[18] K. Tsubakihara et al. (to be submitted)

AUTHOR INDEX

AUTHOR INDEX